AFCRC-TR-56-185

ASTIA Document No.: AD 110170

Radio Propagation Laboratory
Stanford Electronics Laboratories
Stanford University
Stanford, California

METEOR RATE AND RADIANT STUDIES

(Theoretical and experimental radio studies
of meteor ionization trails, with application
to radio propagation by meteor reflections)

by

V. R. Eshleman
P. B. Gallagher
R. F. Mlodnosky

Final Report

February 28, 1957

Prepared under Contract AF 19(604)-1031

The research reported in this document has been sponsored
by the Air Force Cambridge Research Center,
Air Research and Development Command

WILDSIDE PRESS

SUMMARY

From the theoretical and experimental investigations of meteors undertaken under this contract, the following new results have been obtained.

Exact (mode-summing) solutions of the radio-reflecting characteristics of circular columns of ionization reveal that: (1) previous approximate solutions are accurate for low-density trails and parallel polarization (electric vector of incident wave parallel to the trail axis) for scattering in all directions. (2) Single-peak plasma resonance occurs for scattering in all directions from low-density trails with perpendicular polarization of the incident wave. (3) Previous approximate solutions for reflections from high-density trails with parallel polarization are reasonably accurate for all directions of scatter, but the approximate methods are reliable only for back-scatter with perpendicular polarization. (4) In forward scatter from high-density trails, the echo amplitudes and durations are much larger than for back-scatter. (5) The collision frequencies likely to be found in meteor trails are too low to have an appreciable effect on hf and vhf scattering from meteoric ionization. (6) Mode solutions converge so slowly that they are not feasible--even with high-speed computers--when the diameter of the column is much more than one radio wavelength.

An approximate treatment of short-wavelength radio reflections from low-density meteor ionization trails indicates that: (1) for wavelengths shorter than about 3 meters, meteor trails appear to the incident radiation more nearly like paraboloids of revolution than circular cylinders. (2) At short wavelengths, $P_R \sim GA\lambda^4 v^2 D^{-2} R^{-4}$ and $\tau \sim DR\, v^{-2} \lambda^{-1}$, while for long wavelengths, $P_R \sim GA\, \lambda R^{-3}$ and $\tau \sim \lambda^2 D^{-1}$. ($P_R$ = peak echo intensity, G = gain of transmitting antenna, A = aperture of receiving antenna, λ = radio wavelength, v = velocity of the meteoric particle, D = diffusion coefficient, R = radar range, and

τ = echo duration.) The transition between short and long wavelength occurs when $\lambda_T^3 = 2^7 \pi^4 D^2 R v^{-2}$. The echo energy for both long and short wavelengths is proportional to $GA \lambda^3 R^{-3} D^{-1}$. (3) Echo intensity and energy is also greatly affected by the initial radius of the ionized cloud near the meteoric particle. This effect is expected to set a ceiling above which meteor trails cannot be detected by radio, the ceiling being lower at the shorter wavelengths. (4) For meteor studies and meteor propagation, the oblique propagation path geometry combines the advantages of relatively short wavelengths for low receiver noise and high antenna gain with the advantages of an equivalent wavelength several times as long for high echo intensity and small effect of the initial radius.

From a theoretical investigation of meteor trail lengths, it appears that: (1) the modal value of trail length for meteors from the same radiant is 12 sec ζ km for sporadic meteors and 17 sec ζ km for shower meteors, where ζ is the zenith angle of the radiant. (2) For a representative radiant distribution, the modal trail length for sporadic meteors is 15 km. (3) When these sporadic meteors are detected by radio, there is a selection process which favors the detection of the longer trails. For a radar system employing a broad beam antenna, the modal length of the detected sporadic meteors is 25 km, in good agreement with experiment. (4) The length distribution of the trails that are detected is almost independent of the sensitivity of the radio detecting system.

Theoretical studies of meteor rate and radiants show that: (1) more meteors enter an observer's hemisphere during the morning and autumnal equinox than during the evening and vernal equinox periods. The amount of diurnal variation decreases with latitude, while the amount of seasonal variation increases with latitude. (2) A concentration of meteor orbits in the ecliptic plane could produce a shifting in time of occurrence of the morning maximum and evening minimum. (3) There are dead zones, or "anti-showers," in the meteor radiant distribution because of certain restrictions of allowable orbits. (4) for the maximum rate of radar detection of meteors in the northern hemisphere, the antenna beam should be pointed west near the midnight hours, north during the morning hours, east during the noon hours, and south during the evening hours. For an east-west oblique path, the antennas should be pointed north of the great-circle path during the morning hours, and south of this path during the evening hours, for maximum meteoric rate. For maximum meteoric rate over a north-south path, the antenna should be pointed west and east of the direct path during the night and day, respectively. For radars and oblique circuits in the southern hemisphere, north and south should be inter-changed in the above statements.

From radar measurements of the directional properties of meteor reflections, it appears that: (1) the gross directional features predicted from theory are observed. (2) Short term variations in the radiant distribution occur, presumably due to shower-like activity of limited duration and unpredictable time of occurrence. (3) From simultaneous radar and oblique path measurements, it is believed that the apparently-random variations in the radiant distribution could be effectively utilized in meteor propagation over oblique paths by use of an "instantaneous prediction" technique. In this method, information gathered continuously by a radar at one end of an oblique path would be used to determine the optimum antenna bearings for the oblique path.

Measurements of meteor echoes propagated over long paths show that: (1) as path length is progressively increased, echo durations first increase and then decrease, presumably because of the greater mean heights of the meteors which are effective over the longest paths. (2) The dependence of echo duration on radio wavelength is approximately $\lambda^{1.0}$ for obliquely-propagated echoes and wavelengths between about 3.5 and 5.5 meters. This should be compared to a predicted $\lambda^{2.0}$ dependence for long-wavelengths, and a $\lambda^{-1.0}$ dependence for short-wavelengths.

(3) The measured number-intensity distribution of meteor echoes propagated over a very long path is such that the number of echoes of intensity greater than P_R is proportional to $P_R^{-1.0}$. This should be compared to the theoretical $P_R^{-0.5}$ for low-density trails and $P_R^{-2.0}$ for high-density trails. The number-duration distribution of these echoes is similar to the distribution expected at back-scatter.

High-sensitivity radar measurements on very small meteors (15th visual-magnitude, or about 5 visual magnitudes smaller than meteors previously studied by radar techniques) show that the number-intensity distribution is about the same for these meteors as for the relatively large meteors. However, there is some indication that the very smallest meteors may be even more numerous than expected from extrapolation of the number-intensity distribution law determined for the larger meteors.

The results outlined above have application to problems associated with meteor-burst and ionospheric-scatter propagation, and toward a better understanding of the earth's upper atmosphere and the earth's environment in interplanetary space.

ACKNOWLEDGEMENTS

The authors gratefully acknowledge the continued assistance and encouragement of the people at Stanford and at the A.F.C.R.C. who have been connected with this program. In particular, we would mention O. G. Villard, Jr., L. A. Manning, A. M. Peterson, G. H. Keitel, T. V. Harroun, W. Chen, H. Lee, J. Jamieson, G. Durfey, H. Olson, F. Boxall, E. Wilkinson, F. Trinkl, and K. Winkler of Stanford University, A. M. Peterson and W. R. Vincent of the Stanford Research Institute, and P. Newman, J. Casey, J. Aarons, N. Stone, and D. Winter of the A.F.C.R.C.

TABLE OF CONTENTS

LIST OF ILLUSTRATIONS

LIST OF ILLUSTRATIONS (Cont'd)

I. INTRODUCTION

In this final report, the results of the theoretical and experimental investigations of meteors and meteoric ionization obtained on the subject contract are reviewed and discussed. The areas of research which were undertaken are considered in the following order:

(1) Theory of Radio Reflections from Meteor Ionization Trails.

(2) Theoretical Lengths of Meteor Ionization Trails.

(3) Theoretical Meteor Rate and Radiant Determinations.

(4) Radar Measurements of Meteor Ionization Trails.

(5) Properties of Meteor Echoes Propagated over Long Paths.

(6) Measurements of Very Small Meteors.

These topics are the section headings for the body of this report.

The research efforts and presentation of results have been directed both toward a scientific research approach, and toward an application to important new communication techniques based on meteor propagation.

It had been suggested some years ago that meteoric ionization plays an important role in extended-range vhf ionospheric scatter propagation.[1,2,3,4] The amplitude-time characteristics of meteor reflections, however, are not well suited for continuous propagation. An intermittent communication technique using meteor bursts has been pioneered at the Canadian Defense Research Board.[5] In this system, information is sent and received in bursts during the short periods of time for which large, individual meteor trails are providing a reflecting medium for propagation from transmitter to receiver. By using large bandwidths for the small percentage of the time during which information is transmitted, relatively high average rates of information transmission can be achieved.

[1]Superscript numerals refer to the numbered references which appear at the end of the report.

Studies of meteoric phenomena can lead to new information about the solar system, and about our upper atmosphere. Studies of meteors are required to determine the optimum system parameters for meteor burst propagation. Studies of very small meteors are needed to help define the characteristics of ionospheric scatter propagation. The results of some researches in these areas are reported here.

II. THEORY OF RADIO REFLECTIONS FROM METEOR IONIZATION TRAILS

Ideally, in a theoretical development, we would like to have an exact model of the phenomenon under study. In practice, of course, we almost always must be content with exact solutions of an approximate model, or approximate solutions of a more nearly exact model. Often, all we can hope for is an approximate solution of an approximate model.

In the theoretical investigations undertaken under this contract of radio reflections from meteor trails, progress has been made on both exact and approximate treatments of various models of the ionized trail left by meteors. It cannot be stated that any of these models are exact representations of the actual meteor trails. As could be expected, however, it appears that the more nearly correct model can be treated only by approximate methods.

Approximate solutions are often very valuable even when exact solutions have been obtained for the same model. This is especially true when the exact solution requires numerical methods, while the approximate solution yields an explicit expression showing the contribution of each of the variables to the result. Under this condition, the approximate solution can be used to investigate the effects of all possible values of the various parameters, while the exact solution, made for several discrete combinations of the variables, serves as a valuable reference from which the limitations and reliability of the approximate results can be evaluated.

The important results of the theoretical investigations of reflections from meteor trails have been reported in scientific reports and publications. These results are summarized and discussed here.

A. MODE-SUMMING SOLUTIONS

In Scientific Report No. 1 ("Mode Solutions for Radio Waves Scattered by Meteor Trails" by G. H. Keitel, condensed version published in Proc. IRE, 43, 1482, October, 1955) exact evaluations are presented of radio reflections from infinitely-long, circularly-symmetric columns of ionization. The method used to determine the reflection coefficients may be applied to any circularly-symmetric distribution of ionization, though the model chosen for numerical computations is a column in which the radial distribution of ionization density is a Gaussian function. (It is believed that this distribution is a good approximation to the actual meteor trail, since it would result from ionization diffusing uniformly from an instantaneous line source.)

The differential equations which describe the electro-magnetic fields within the column of ionization for a normally-incident wave were integrated numerically on the National Bureau of Standards Western Automatic Computer (SWAC) at the University of California at Los Angeles. For each assumed value of linear ionization density, column size, and collisional loss, the significant mode reflection coefficients were cal-culated at three matching radii. (Since the electron volume density was assumed to vary as the Gaussian function with radius, there is no definite boundary to the column.) A comparison shows that the size of the matching radius has a negligible effect, providing the electron volume density at the matching radius is sufficiently low.

The most extensive sets of computations were performed for two values of the electron line density. A line density of 10^{13} electrons per meter was chosen to compare with approximate theories of reflection from low-density (line density $q < 10^{14}$ electrons per meter) trails. A line density of 10^{17} electrons per meter was chosen to compare with approximate theories of reflections from high-density trails. The principal results of these numerical computations are presented in Figs. 1-4. In each figure, the magnitude of the reflection coefficient ρ is plotted on polar coordinates for the various angles of scatter from the trail. The trail is at the origin, and the incident radiation is traveling in the $\theta = 0^\circ$ direction. The various curves on each plot are for different values of kr_o, where k is the wave number $2\pi/\lambda$, λ is the radio wavelength, and $r_o/\sqrt{2}$ is the inflection radius of the Gaussian distribution of ionization volume density. For normal diffusion, $r_o^2 = 4$ Dt, where D is the diffusion coefficient and t is the time elapsed since the formation of the line of ionization. The magnitude of the reflection coefficient, as defined for cylinders, is related to other parameters by

$$\rho^2 \equiv \frac{k}{4R}\ \sigma \equiv \frac{4\pi k^3 R^3}{G^2}\ \frac{P_R}{P_T} \tag{1}$$

where R is the radar range, σ is the standard radar cross section, G is the antenna power gain over an isotropic radiator, P_R is the echo power, and P_T is the transmitted power.

In Fig. 1 ρ is plotted as a function of θ for parametric changes in (kr_o), and for $q = 10^{13}$ electrons per meter and parallel polarization (electric vector of the incident wave parallel to the axis of the ionization column). In Fig. 2 the same parameters are considered, except that the results are for perpendicular polarization. Comparison of the results pictured in Fig. 1 with the approximate "scatter" theory for

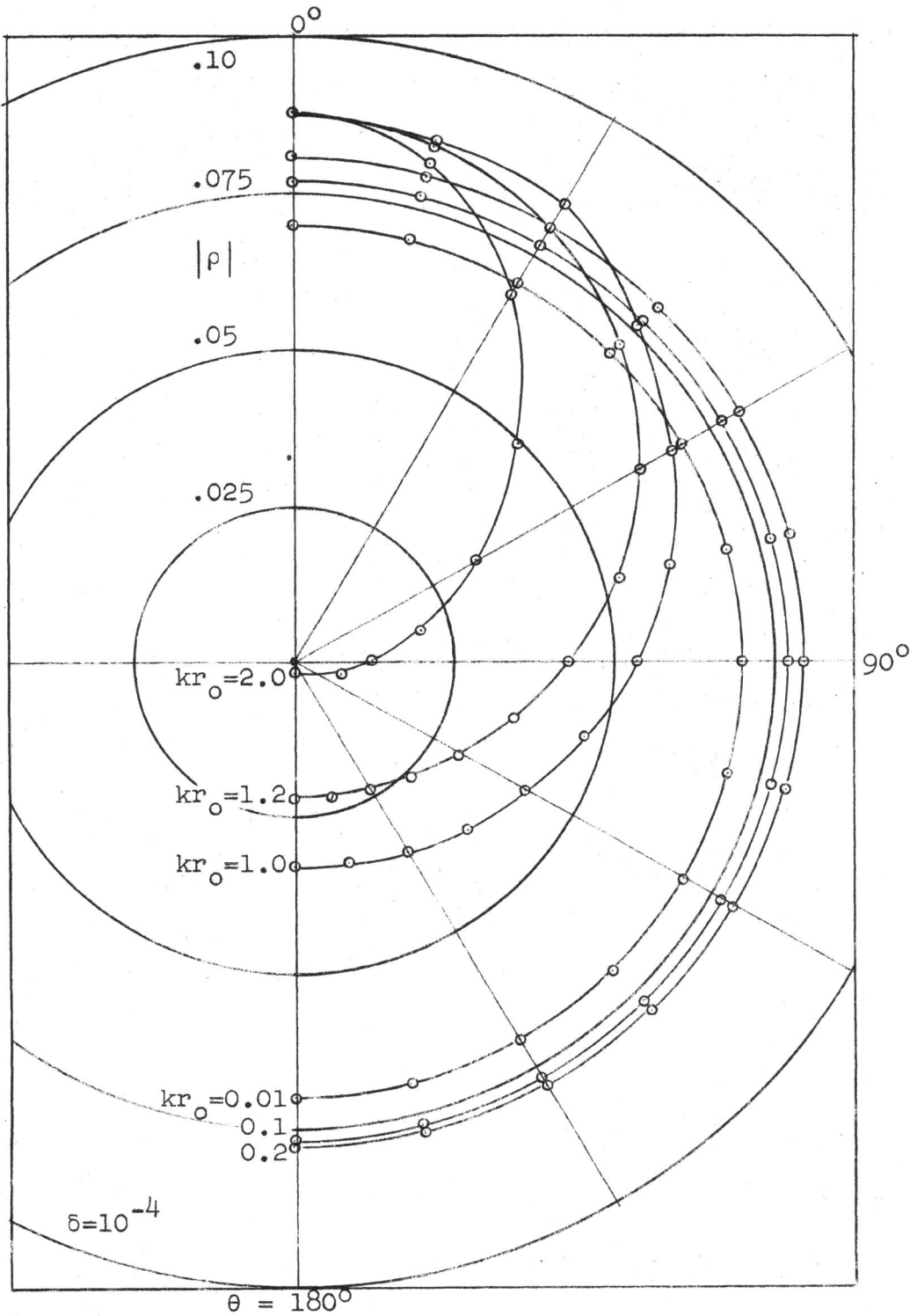

FIG. 1.—Parallel reflection coefficient polar diagram, $q = 10^{13}$ electrons per meter.

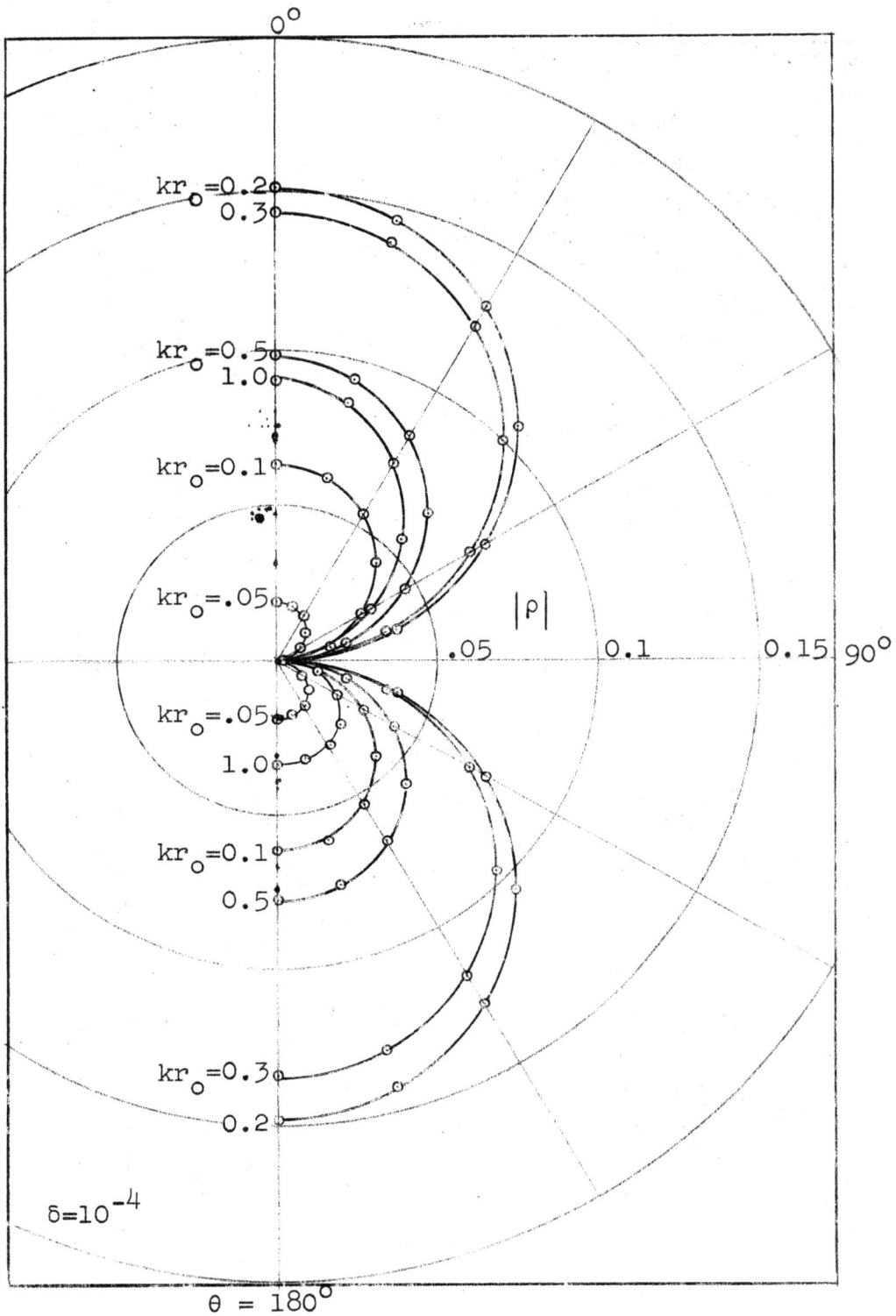

FIG. 2.--Transverse reflection coefficient polar diagram,
$q = 10^{13}$ electrons per meter.

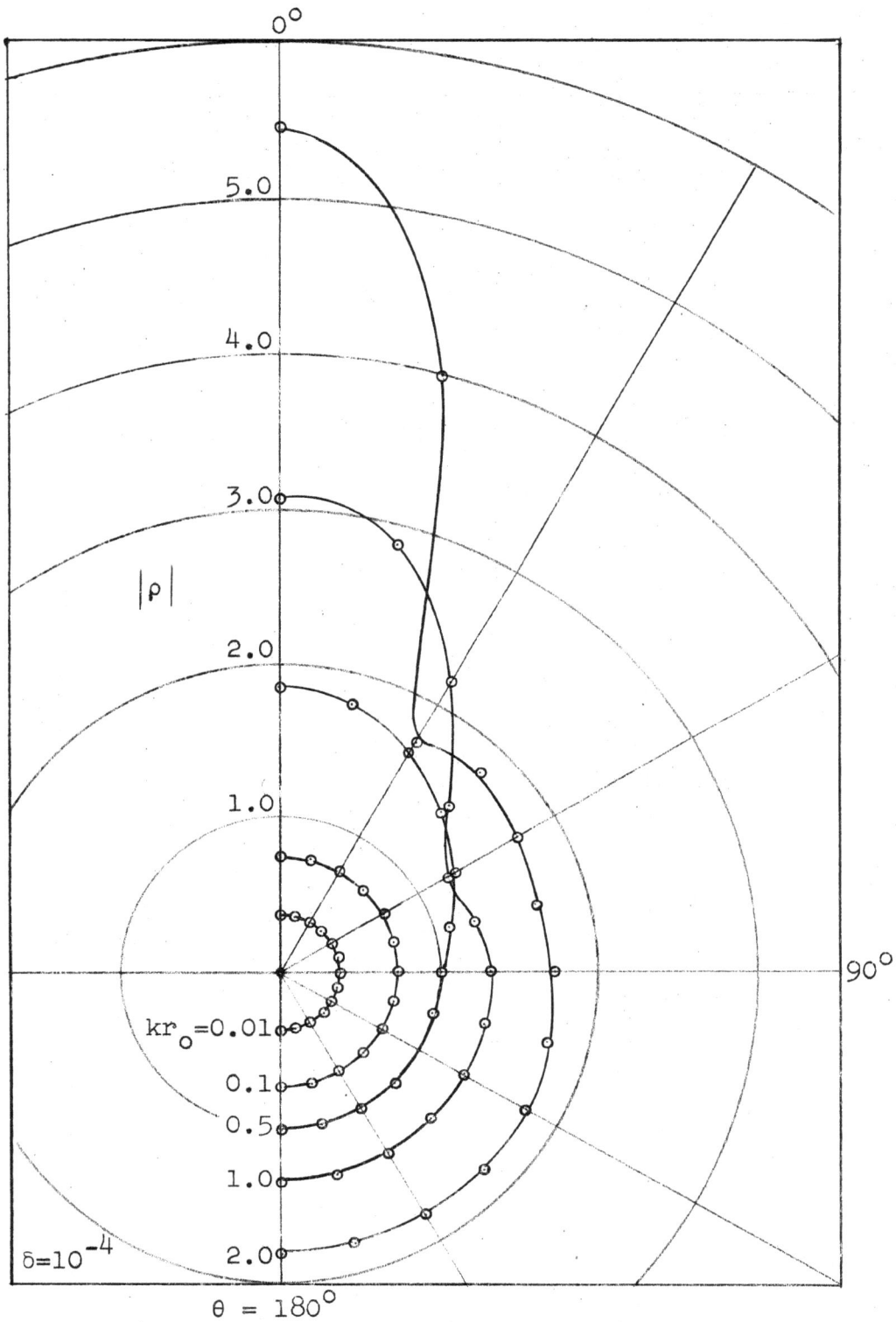

FIG. 3.--Parallel reflection coefficient polar diagram,
q = 10^{17} electrons per meter.

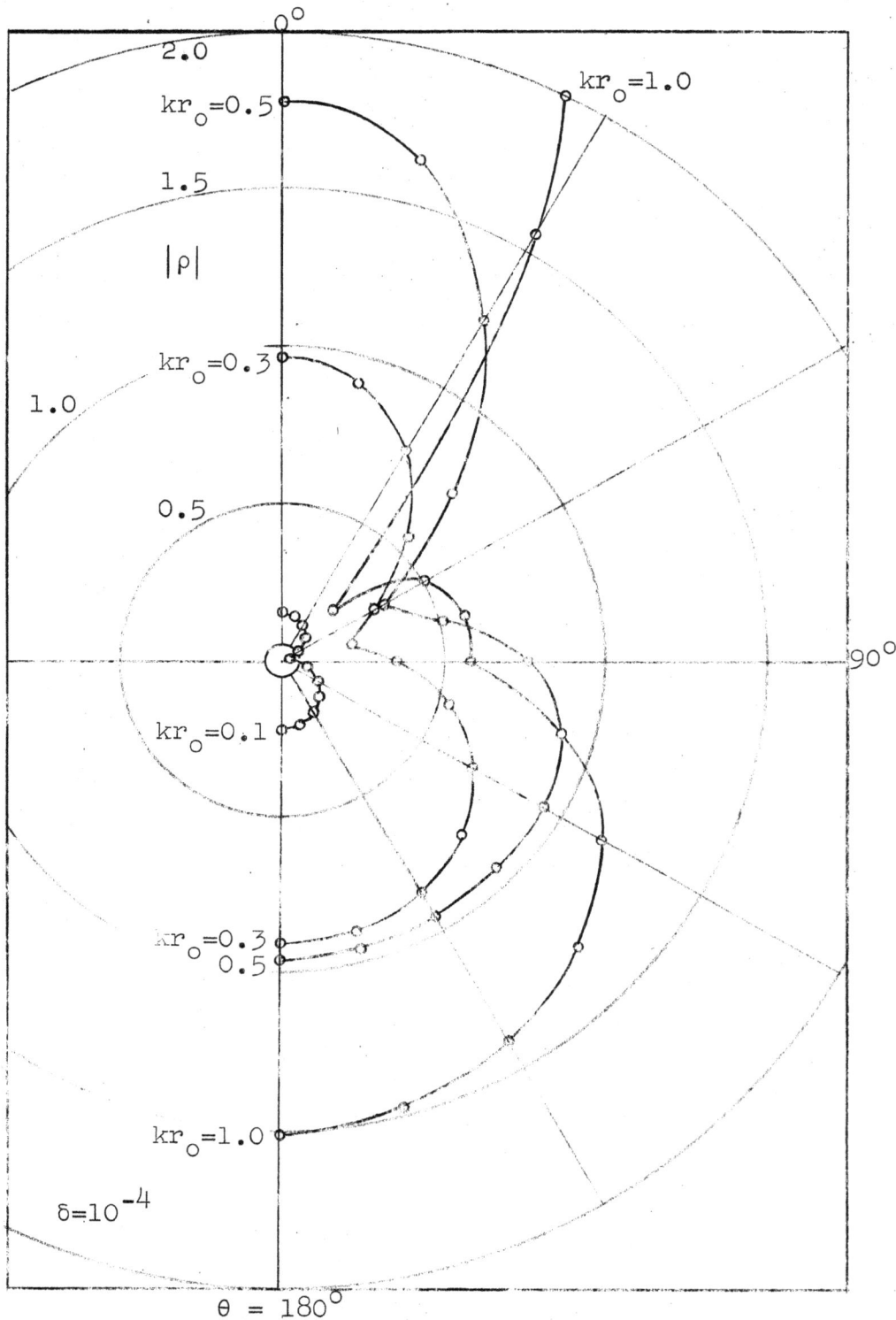

FIG. 4.--Transverse reflection coefficient polar diagram, $q = 10^{17}$ electrons per meter.

scatter in all directions from low-density trails[6] indicates
that the approximate theory gives excellent agreement for the
parameters considered. Comparison of the results pictured in
Fig. 2 with the approximate theory indicates that the approximate theory is good for trail sizes larger than the "resonant"
size,[7,8] but not for trails equal to this size and smaller.
A new result of the exact theory is the demonstration that
plasma resonance occurs for all angles of scatter from a low-
density trail. For application to meteor burst propagation,
perhaps the most important result of the exact computations
for low-density trails is the verification of the dependence
of echo duration on scattering angle. That is, as predicted
from the approximate theory,[6] it was found in the exact
computations that the duration of the short-duration echoes
from low-density trails is proportional to $\sec^2\phi$, where 2ϕ
is the angle between the rays from the meteor trail to the
transmitter and receiver.

Figures 3 and 4 present values of ρ for $q = 10^{17}$ electrons
per meter and parallel and perpendicular polarization, respectively. Of particular significance is the growth of the geometrical optics shadow lobe (the large lobe toward $\theta = 0°$) for
large values of (kr_o). Unfortunately, it was not feasible to
compute the reflection coefficients for larger values of
cylinder size of the high-density trail because of the slow
convergence of the series solution. However, the limited
results pictured in the figures are sufficient to show that
forward-propagated echoes from high-density meteor trails are
larger in amplitude and greater in duration than the back-
reflected echoes. These results may help explain the discrepancy between echo rate-amplitude distributions measured
by radar and oblique-path techniques. In these measurements
there are larger percentages of large-amplitude echoes recorded
on the oblique paths than in the radar measurements.

In the published approximate theories for scattering from high-density trails, results are presented only for the back-scattered echoes. A comparison of the exact results with approximate theory shows reasonably good agreement for $\theta = 180^{\circ}$. An unexpected result of the comparison is the apparent close agreement between the exact and approximate treatments for all values of θ for parallel polarization, and the lack of agreement for values of θ other than 180° for perpendicular polarization. The explanation appears to be in the differences in current flow in the actual ionization column and in the metallic cylinder used in the approximate theory.

In the computed mode solutions, only the case of a plane wave incident normally onto the column was considered. For general incidence onto a non-homogeneous column, it is no longer possible to separate the problem into independent considerations of the parallel and perpendicular polarization cases. Coupling between the two polarizations exists throughout the column. The formulas for the general case were developed, but no numerical computations were attempted. In the case of the low-density trails, however, the favorable comparison of the two theoretical approaches adds confidence to our reliance on the approximate theory for general angles of incidence and reflection.

In order to perform the numerical integrations for perpendicular polarization, it was necessary to introduce collisional losses to avoid a singularity in the function. Computations using various values of collision frequency, and considerations of the likely values encountered in the meteor trail, indicate that collisional loss is probably not an important factor in the magnitude of meteor reflections.

In Scientific Report No. 2 (The Program of a Two-Dimensional Scattering Problem for an Automatic-Sequenced Digital Computer by G. H. Keitel), the details of the numerical integration program of the SWAC computer for the meteor-trail reflection problem are presented. The numerical integration program which was used is based on Milne's three point method, and has the special feature of automatically halving or doubling the integration step size when the conditions of the problem require or permit. For this particular problem, the halving and doubling feature resulted in a considerable saving in computer time. Further considerations of Milne's three-point method and the variable step size are presented in Scientific Report No. 6 ("An Extension of Milne's Three-Point Method," by G. H. Keitel, also published in the Journal of the Association for Computing Machinery, 3, 212; July, 1956).

In Scientific Report No. 7 (On the Dipole Resonant Mode of an Ionized Gas Column," by G. H. Keitel, also published in the Australian J. of Phys., 9, 144; 1956) a detailed investigation of the plasma resonance peak for perpendicular polarization is presented. This work was instigated by the suggestion by Makinson and Slade[9] that more than one resonance should be observed as a meteor trail expands. In the detailed computations of plasma resonance, however, it was apparent that only one peak could exist. It is believed that the multiple peaks obtained by Makinson and Slade were produced by the discontinuities of electron density in their stepped approximation to a Gaussian distribution of electron density in the trail.

B. SCATTER-THEORY SOLUTIONS

In Scientific Report No. 5 ("Short-Wavelength Radio Reflections from Meteoric Ionization: Part I; Theory for Low-Density Trails," by V. R. Eshleman) the approximate theory of scattering from low-density trails is expanded to include the conditions where the differential radial size of the column (resulting from the finite velocity of the meteoric particle) must be taken into account. The exact computations discussed above were for columns having translational and circular symmetry. The assumption of translational symmetry cannot apply, however, when the time required for the meteoric particle

to produce the specularly-reflecting region (the principal
Fresnel zone) is comparable to the time for the low-density
column to expand to a size where it no longer reflects
appreciable energy. But this condition occurs for the shorter
radio wavelengths, since the transit time of the particle
through the principal Fresnel zone is proportional to $\lambda^{1/2}$,
while the trail expansion time referred to above, is proportional
to λ^2. Thus at the shorter wavelengths, it is important to
consider the ionization trail as the paraboloid of revolution
produced by a moving point-source of ionization. We have not
made any attempts to solve for reflections from such a trail
model by the exact, wave-matching technique. For the low
density trails, however, this geometry is not much more difficult
to treat than circular columns with translational symmetry.

In Tables I and II are summarized the analytical solutions
for the intensities and durations of meteor echoes obtained
from the approximate theories. For low-density trails, the
results were obtained from the scatter theory, where it is assumed
that each free electron is acted upon by the incident wave
only, and that the scatter field is the vector sum of the
fields scattered from each electron. At long wavelengths, the
trail is assumed to be a long cylinder of constant diameter,
while at short wavelengths the model discussed above is employed.
For high-density trails and long wavelengths, it is assumed
that total reflection occurs at the surface where the electron
volume density is critical (i.e., where the equivalent dielectric
constant is zero). Solutions have not yet been obtained for
high densities and short wavelengths.

The symbols occurring in Tables I and II are identified
below:

λ radar wavelength, meters

λ_T transitional wavelength, equal to
$2^{7/3}\pi^{4/3}D^{2/3}R^{1/3}v^{-2/3}$ meters

D diffusion coefficient, square meters per second

R radar range, meters

TABLE I: Peak echo intensities P_T (Watts).

	Long wavelengths $\lambda > \lambda_T$
Low Density $q < 10^{14}$	$P_T \text{ I } q^2 \dfrac{\lambda R}{2} \exp\left(-\dfrac{8\pi^2 r_i^2}{\lambda^2}\right)$
High Density $q > 10^{14}$	$P_T \text{ I } q^{1/2} \dfrac{\lambda R}{4\pi \exp(0.5)\left(\dfrac{\mu_o e^2}{4\pi m}\right)^{3/2}}$
	Short wavelengths $\lambda < \lambda_T$
Low Density $q < 10^{14}$	$P_T \text{ I } q^2 \left(\dfrac{v \, \lambda^2}{16\pi^2 D}\right)^2 \exp\left(-\dfrac{8\pi^2 r_i^2}{\lambda^2}\right)$
High Density $q > 10^{14}$	————————

- 13 -

TABLE II: Intrinsic echo durations τ (seconds).

	Long Wavelengths $\lambda > \lambda_T$
Low Density $q < 10^{14}$	$$\frac{\lambda^2}{32\pi^2\,D}$$
High Density $q > 10^{14}$	$$\frac{\lambda^2}{4\pi^2\,D}\ \left(\frac{\mu_o e^2}{4\pi m}\right)\ q$$
	Short Wavelengths $\lambda < \lambda_T$
Low Density $q < 10^{14}$	$$\frac{4\pi^2\,D\,R}{v^2\,\lambda}$$
High Density $\lambda > 10^{14}$	————————

v	meteor velocity, meters per second
q	electron line-density, electrons per meter
P_T	transmitted peak power, watts
I	received intensity per unit transmitter power for one free electron, equal to $G_T G_R \lambda^2 (4\pi)^{-2} R^{-4} (\mu_o e^2/4\pi m)^2$
G_T, G_R	power gain over isotropic antennas of the transmitting and receiving antennas.
$\dfrac{\mu_o e^2}{4\pi m}$	the classical electron radius, 2.8178×10^{-15} meters. Here μ_o is the permeability of free space, and e and m are the electron's charge and mass, respectively
r_i	initial radius of the cloud of ionization formed near the meteoric particle, meters

The expressions in the tables are for radar detection of meteor trails, where the transmitter and receiver are at the same location. For meteor burst propagation over oblique paths, the low-density formulas in the tables are applicable when the following changes are made:

(1) Replace λ by $\lambda \sec \emptyset$ everywhere in the tables.

(2) Replace R by $M(1-\sin^2\emptyset \cos^2\beta)^{-1}$ everywhere in the tables.

(3) Note that I for oblique propagation is equal to $G_T G_R \lambda^2 \sin^2\alpha (4\pi)^{-2} R_1^{-2} R_2^{-2} (\mu_o e^2/4\pi m)^2$.

In these last expressions, the symbols are:

R_1, R_2	ranges from the transmitter and receiver to the meteor trail, meters.
α	angle between the electric vector at the meteor trail and R_2
$2\emptyset$	included angle between R_1 and R_2
β	angle between the meteor trail axis and the plane determined by R_1 and R_2
	$M = 2R_1 R_2 \cos \emptyset (R_1 + R_2)^{-1}$ = the length of the bisector of the angle $2\emptyset$ in the triangle formed by R_1, R_2, and a straight base line, meters.

We know of no simple yet reliable method for changing the back-scatter formulas for high-density trails so that they would apply to propagation over an oblique path.

The intensity-time variations of the meteor echoes as deduced from the approximate formulas are illustrated in Fig. 5. The intrinsic durations τ are indicated on the echoes. For the high-density, long-wavelength case, τ is the total duration of the echo. For low density and long wavelengths, τ is the time the echo exceeds $\exp(-1) = 36.8\%$ of its maximum intensity. For low density and short wavelengths, τ is the time the echo exceeds $(1 + \pi^2/4)^{-1} = 28.8\%$ of its maximum intensity. It is interesting to note that for the low density trails, the product of echo intensity and intrinsic duration (echo energy) is the same for long and short wavelengths.

The durations d of the low-density echoes above a fixed reference level are related to the intrinsic durations by $d(\text{long-}\lambda) = \tau\ (\text{long-}\lambda)\ \ln(q/q_o)^2$ and $d(\text{short-}\lambda) = \tau(\text{short-}\lambda)\ 2[(q/q_o)^2 - 1]/\pi$. In these expressions, q is the line density and q_o is the line density that this trail would have if its peak echo intensity were equal to the reference level.

The expression for the transitional wavelength between long- and short-wavelength theory was given in the list of symbols for the tables. Some discussion is in order concerning the expected value of λ_T. For general radar or oblique prop- agation, the factors which affect λ_T are the geometry of the propagation path from transmitter to receiver, the orientation of the trail relative to this path, the height of the meteor trail (the height determines the diffusion coefficient), and the velocity of the meteoric particle. Assuming nominal values for these parameters, λ_T is near 3 meters for backscatter, 1.5 meters for trails which are parallel to a long, oblique path, and 0.5 meters for trails which are perpendicular to a long oblique path. Under extreme conditions of velocity, height, and path geometry, λ_T may be as great as 100 meters or as short as 0.15 meters.

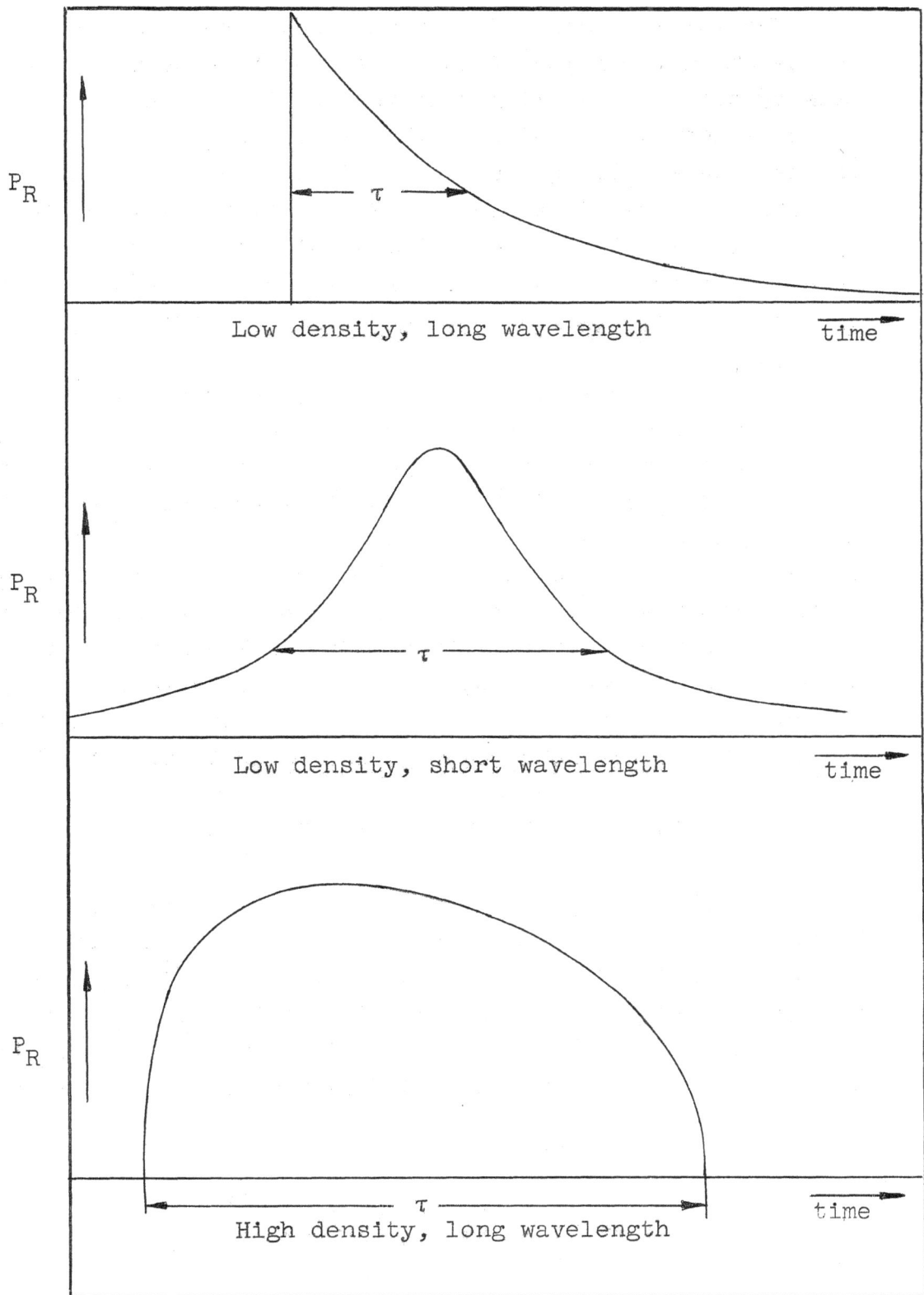

FIG. 5.--Echo shapes determined from the approximate theories.

The initial-radius factor may play a very important role in the characteristics of short-wavelength echoes from low-density trails. Assuming that the initial radius is equal to the molecular mean-free-path, it appears that the initial radius sets a ceiling above which it is virtually impossible to detect a low-density trail. This ceiling is lower for the shorter wavelengths. The initial-radius factor may be so severe that only high-density trails can be detected at the shorter wavelengths, even though high transmitter power and high-gain antennas are used.

All of the characteristics of meteor echoes at short wavelengths are potentially important in applications to meteor burst propagation since this type of propagation is expected to be useful not only in the 10 to 5 meter range, but also for wavelengths as short as 3 meters, or even shorter. It appears important to continue the theoretical work on short-wavelength characteristics of meteor echoes, and to try to complete the approximate treatment by a theory for high-density trails in oblique propagation at long and short wavelengths.

Other discussions of the theory of short-wavelength meteor reflections are contained in references 10 through 14.

III. THEORETICAL LENGTHS OF METEOR IONIZATION TRAILS

In Scientific Report No. 4, ("The Theoretical Length Distribution of Ionized Meteor Trails" by V. R. Eshleman, also published in J. Atmosph. Terr. Phys., Feb., 1957), the lengths of ionized meteor trails are determined from theoretical and experimental information on the production and radio detection of meteoric ionization. An approximate formula for the trail length L is

$$L \cong \frac{12}{7} \, (H \sec \zeta) \left[2 \ln \left(\frac{q_{max}}{q_{min}} \right) \right]^{\frac{1}{2}} \qquad (2)$$

where H is the scale height (approximately 7×10^3 meters), ζ is the zenith angle of the meteor radiant (direction of the velocity vector), q_{max} is the maximum electron line density along the trail, and q_{min} is the value of line density chosen to represent the ends of the trail. (For example, if q_{min} is the minimum detectable line density for a particular system, then L is the length of the trail as far as that system is concerned.) Theoretically, q_{max} is proportional to the mass of the meteoric particle and to the cosine of the zenith angle. In Fig. 6, the trail length given by (2) is plotted as a function of zenith angle and meteor mass.

The distribution of meteor trail lengths can be determined from the number-mass distribution of the meteoric particles, and the distribution of meteor radiants. The most likely (model) value of trail length for sporadic meteors from the same radiant is found to be 12 sec ζ km. The distribution of sporadic trail lengths is relatively insensitive to the exact radiant distribution. For a representative radiant distribution (radiant density varying as the sine of the zenith angle), the most likely trail length for sporadic meteors is 15 km.

When meteor trails are detected by a radar system, there is a selection process which favors the detection of the longer trails. The curve of the theoretical trail-length distribution for meteors detected by a radar employing a broadbeam antenna is compared with the experimental results of Manning, Villard, and Peterson[15] in Fig. 7. The most likely trail length of the detected trails in both theory and experiment is approximately 25 km. The theory indicates that the length distribution is almost independent of the sensitivity of the radar receiver, or the power output of the radar transmitter.

Considerations of trail length play an important role in studies of the scattering polar diagram of meteor trails, the fading of long-enduring meteor echoes, and the ratio of the number of meteors detected to the total number of meteors.

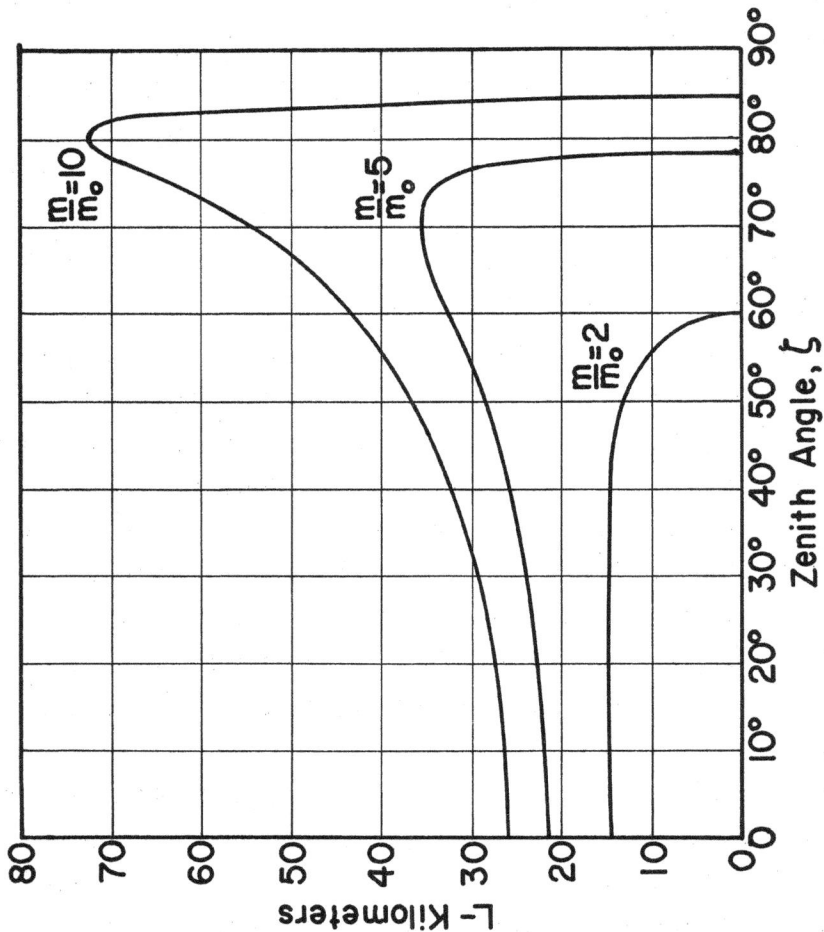

FIG. 6.--The length L of meteor trails plotted as a function of particle mass m and zenith angle of approach ζ
into the upper atmosphere. Meteor trail length is defined
here as the distance along the ionized trail between two
positions where the line density q is a fixed value, equal
to q_{min}, where $q_{min} = q_{max}$ when $m = m_o$ and $\zeta = 0$.

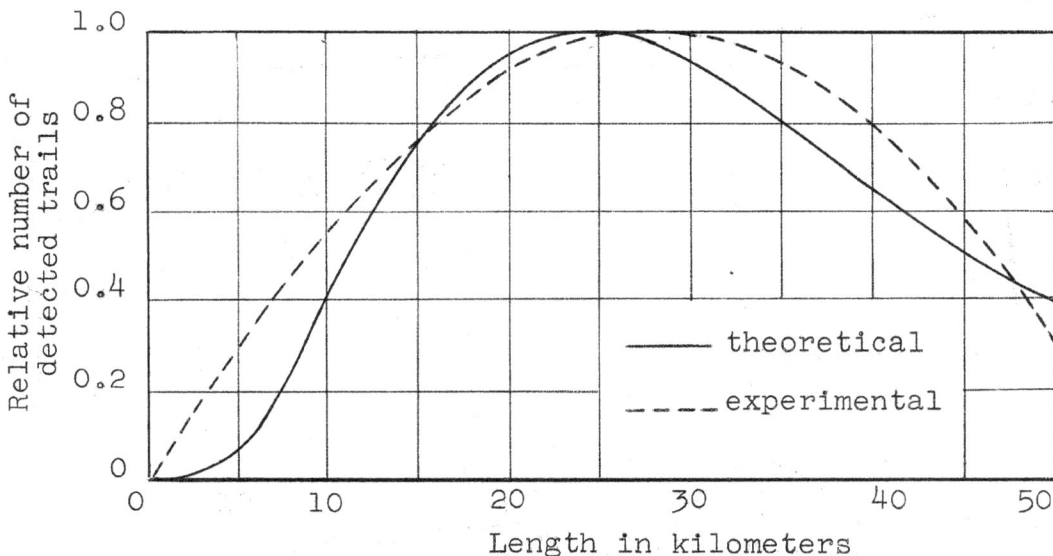

FIG. 7.--Relative number of sporadic trails detected, plotted as a function of their length. For the theoretical curve, it is assumed that the area distribution of sporadic radiants over the celestial hemisphere is proportional the sine of the zenith angle, and that the gain function of the radar antenna is proportional to the sine of the elevation angle. The experimental curve is from Manning, Villard, and Peterson.[15]

IV. THEORETICAL METEOR RATE AND RADIANT DISTRIBUTIONS

A. TOTAL RATES

In Quarterly Status Report No. 2 (August 15, 1954) were presented the results of an investigation of two simple theoretical models for the distribution of meteor orbits in space. By considering the manner in which the earth moves through this matter, the number and directions of arrival (radiants) of the meteors were determined for several locations on the earth. The results for the total rates are repeated in Figs. 8 through 11. Figures 8 and 9 show the diurnal and seasonal variations of meteor occurrence rate assuming a spherically-uniform heliocentric distribution of meteor orbits. It is also assumed that all of the meteors are moving at the parabolic limit velocity of 42 km/sec with reference to the sun.

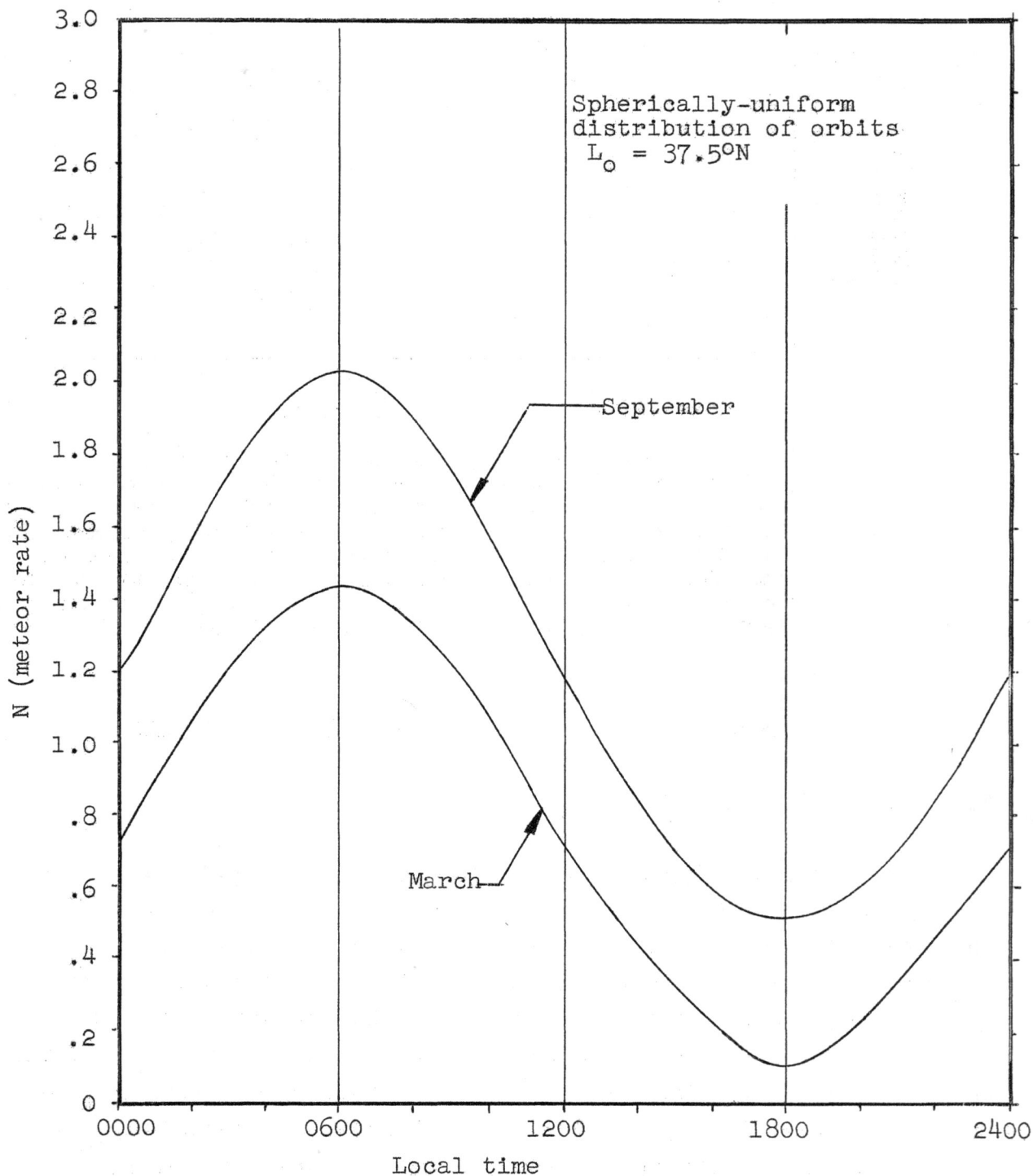

FIG. 8.--Theoretical diurnal and seasonal variation of meteor rate at 37.5°N latitude. Meteors are assumed to be traveling in random heliocentric directions at the parabolic limit velocity.

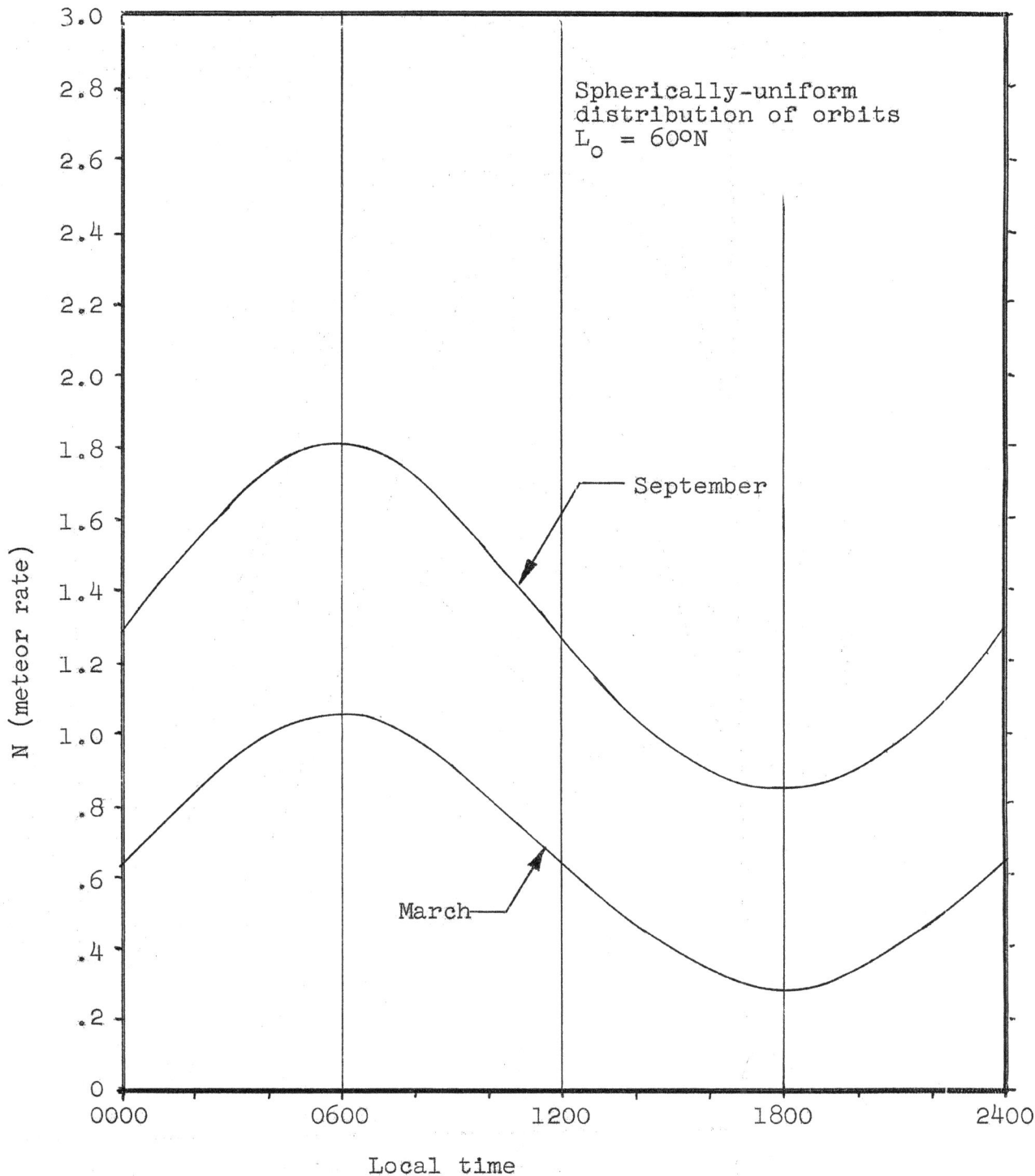

FIG. 9.--Theoretical diurnal and seasonal variation
of meteor rate at 60°N latitude. Meteors are assumed to
be traveling in random heliocentric directions at the
parabolic limit velocity.

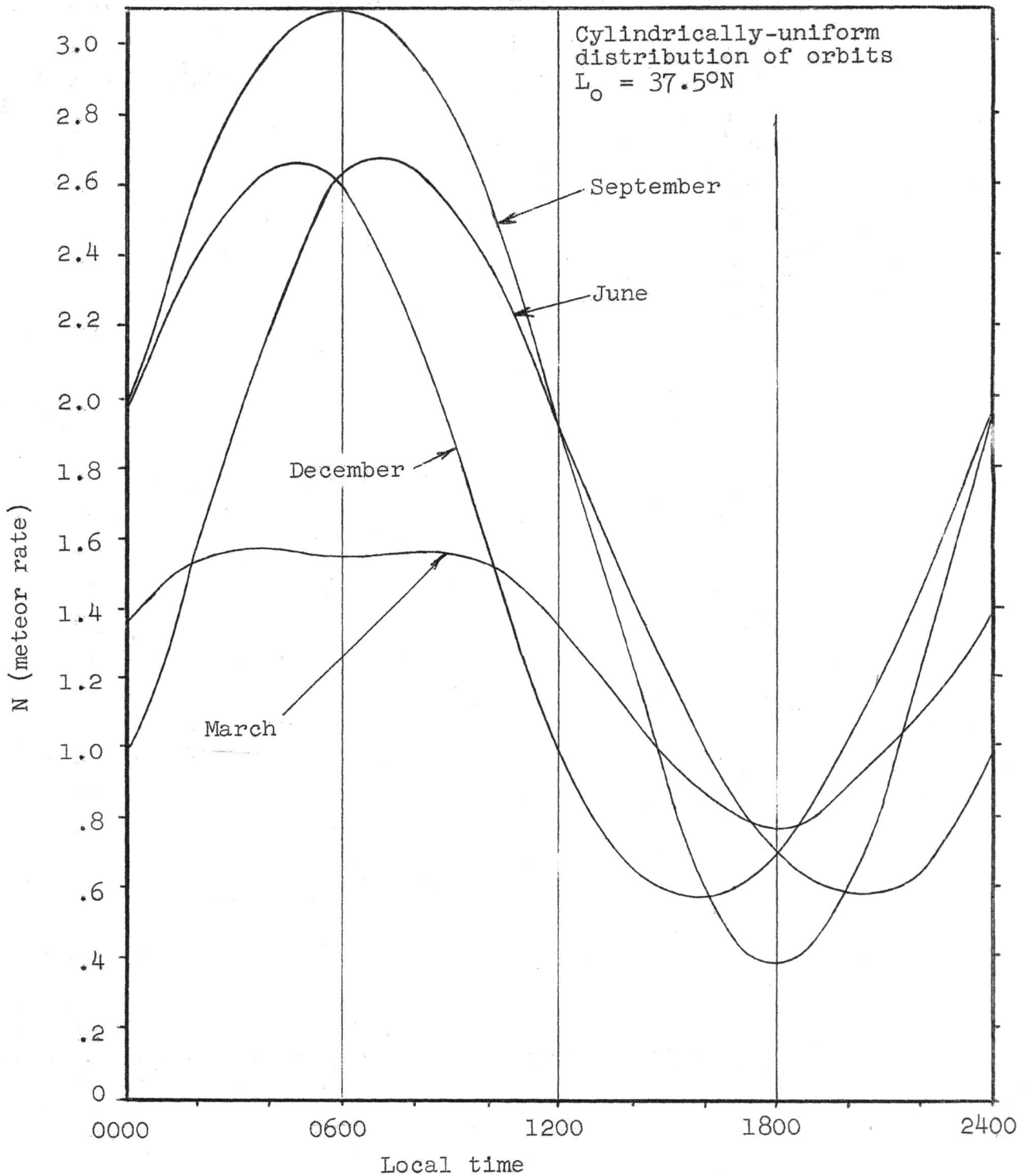

FIG. 10.—Theoretical diurnal and seasonal variation of meteor rate at 37.5°N latitude. Meteor orbits are assumed to be in the plane of the ecliptic, and - within this plane - they are assumed to have random heliocentric directions and to be traveling at the parabolic limit velocity.

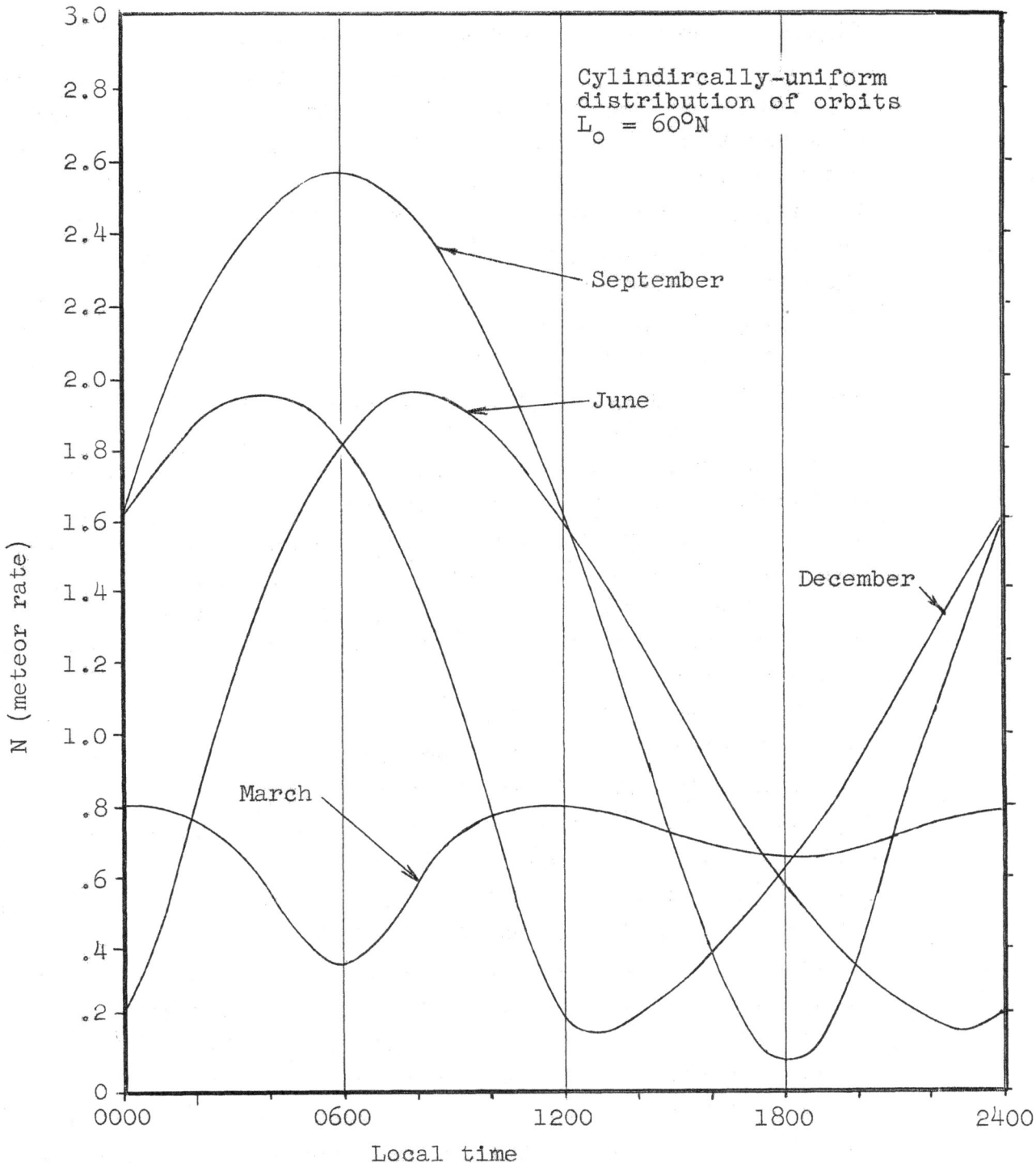

FIG. 11.--Theoretical diurnal and seasonal variation of meteor rate at 60°N latitude. Meteor orbits are assumed to be in the plane of the ecliptic, and - within this plane - they are assumed to have random heliocentric directions and to be traveling at the parabolic limit velocity.

Figure 8 pertains to an observer at latitude $37.5^\circ N$ (the approximate latitude of Stanford University). The abscissa is the local time, and the ordinate is the normalized number of meteors entering the celestial hemisphere of the observer per unit time. The value of unity for the ordinate corresponds to the rate which would be observed at a point in space which is not moving relative to the sun. Only the curves for March and September (the vernal and autumnal equinoxes) are shown. The values for the rest of the year will lie between these two curves, with the variation in June and December being about midway between the plotted curves. The morning (0600) maximum and evening (1800) minimum of meteor rate have long been observed, and are predicted from the simplest geometrical considerations. The seasonal variation, with the maximum rate in the second half of the year, results from the seasonal variation of the minimum angle between the observer's zenith and the apex of the earth's way. These curves apply to the same southern latitude ($37.5^\circ S$) when the March and September curves are interchanged.

In Figure 9, the observer's latitude is taken as $60^\circ N$ (or $60^\circ S$, with the proper change in date). It is seen that as the observer's latitude is increased, the maximum diurnal variation in rate decreases, while the amount of the seasonal variation in rate increases.

Figures 10 and 11 again present the diurnal and seasonal rate changes at $37.5^\circ N$ and $60^\circ N$ latitude, but here it is assumed that the meteor orbits have a cylindrically uniform heliocentric distribution. That is, it is assumed that meteor orbits are in the plane of the ecliptic, the meteors all have the same velocity (equal to the parabolic limit), and, within the plane of the ecliptic, they have random heliocentric directions. An interesting anomaly appears in these curves. While the rate variations for June and December (the summer and winter solstices) are identical when a spherical distribution of meteor orbits is assumed, the times of occurrence

of the maximum and minimum points are markedly different in
the case of the cylindrical distribution. In particular, the
June maximum occurs after 0600, while the December maximum
occurs before 0600, the amount of the deviation from 0600 increasing with the latitude of the observer. Similarly,
the minimum points occur at times considerably different from
the time (1800) of occurrence of the minimum for the spherical
distribution of orbits. In September, the maximum and minimum
of the diurnal rate variations are at 0600 and 1800; however,
it should be observed that the amount of diurnal variation does
not decrease with increasing latitude as much as it did for
the spherical distribution. In March there is an interesting
splitting of the maximum, giving two peaks to the curve which
are equally spaced in time from 0600. In fact, at $60^{\circ}N$
latitude, this effect is so severe in March that there is a
minimum of meteoric activity at 0600, with maxima at noon and
midnight.

The qualitative results of the theoretical rate study
with the meteor orbits confined to the plane of the ecliptic
agree with experimental radar results published by Eastwood
and Mercer.[16] Their measurements, obtained at about $55^{\circ}N$
latitude, show three main features - a late afternoon minimum
and an early and late morning maximum, with the relative
amplitudes of the maxima changing throughout the year. In
December, the early morning maximum is predominant, while in
June only the late morning maximum is observed. In March, both
morning maxima are obtained, while in September, only one broad
maximum occurs, centered near 0600. While these same features
are prominent in the curves presented in Fig. 11, it is important
to note that the theoretical results were for all meteors
entering the observer's celestial hemisphere, while the radar
results apply only to those areas of the sky to which the
radar antenna is sensitive. This factor is considered in
greater detail in the next section.

While the above results help explain the source of some of the gross features of the diurnal changes in meteor rate, the models of the orbital distribution of meteors which were employed in the computations are much too idealized. From radio investigations of meteors, made at the University of Manchester, Hawkins[17] found that the vast majority of meteors travel around the sun in the same direction as the earth. Hawkins also demonstrated that the conversion from heliocentric to geocentric reference, used previously by him and also used in the computations for Figs. 8 to 11, is in error. His new conversion factor causes a greater concentration of radiants toward the apex of the earth's way due to the 30 km/sec orbital motion of the earth.

To include the demonstrated preponderance of direct to retrograde orbits, a new model of the distribution of meteor orbits was considered. In this model, the number density of meteors arriving at the earth from the various heliocentric directions is described by an ellipse of revolution about the apex-antapex line, with the earth at the focus nearest the apex direction. The eccentricity of this ellipse was chosen as 0.9, so that the theoretical heliocentric distribution appears similar to Hawkins's results for the longitudinal distribution of orbits. The revised factor for changing from the heliocentric to geocentric reference was then employed in this distribution. This revised correction factor was also employed on the models considered in Figs. 8 and 9. Another model which was investigated is similar to the ellipsoid described above, with an added correction for the possible dependence of ionization density on meteor velocity. In this model, the maximum electron line density is assumed to be proportional to v^3. The results from these new models are in many respects similar to those pictured in Figs. 8 and 9, the most striking difference being in the ratio of the diurnal maximum to minimum rates. The results for this ratio are summarized in Table III.

TABLE III: Ratios of duirnal mix to mix meteor rates.

Distribution	Vernal Equinox Max/Min	Autumnal Equinox Max/Min
Uniform distribution considered in Figs. 8 and 9	14.2	4.6
Uniform distribution with new correction factor	20	6.7
Ellipsoidal distribution	3.5	2.4
Ellipsoidal distribution with velocity factor	45	9.9

Perhaps the most important conclusion to be gained from the investigations leading up to the results presented in Table III is the observed interplay of the effects of the distribution of meteor orbits and the dependence of line density on velocity. It appears that a separation of these effects from observations of meteor rates and radiants would not be feasible. If the velocity dependence can be determined reliably from other data, a separation of these factors can be made. General agreement on the magnitude of the velocity effect is lacking. From the observed max/min ratios of meteor echoes and the considerations outlined above, however, it can be concluded that if there is a strong velocity effect, there must be many more direct-moving, ecliptic-plane-concentrated, short-period (approximately one year) meteors than previously have been suggested. Such a concentration of orbits would also help explain the apparent discrepancy between the observed number of meteors entering the upper atmosphere, and the number of small particles needed in space to account for the zodiacal

light and the gegenshein. That is, there may be vast numbers
of meteoric particles in space, but because they are moving
nearly in phase with the earth, relatively few collide with us.

In an attempt to deduce further characteristics of the
distribution of meteor orbits, a mathematical analysis was
made of all possible orbits intercepting the earth. The
analysis was made under the following conditions: (1) an
assumed circular orbit for the earth, (2) the earth considered
stationary, (3) effects of earth's gravitational field not
considered, and (4) other astronomical effects and/or pertur-
bations not considered.

From a physical consideration of the finite size and
dissipative power of the sun, certain dead spots are implied
from which no meteors should appear to radiate. The mass of
the meteor and the dissipative power of the sun will determine
a minimum distance at which such a particle can exist from the
sun. Applying these restrictions to the perihelions of the
above orbits, two conical surfaces of revolution have been
indicated as dead spots. Their locations are on both sides
of the earth through an axis along the line joining the sun and
the earth. For a limiting perihelion of 0.05 times the
distance from the earth to the sun (approximately ten diameters
of the sun), half the internal angle of this cone is 12 degrees
for a heliocentric velocity of 42 km/sec. There is a great
velocity dependency of this half-angle. For the same limiting
perihelion of 0.05 and a velocity of 30 km/sec, the angle is
18 degrees. The curves of velocity vs angle with limiting
perihelion as a parameter are shown in Fig. 12. Given a
limiting perihelion as a function of meteor mass size, only the
velocities and directions of radiants are possible which lie
above and to the right of the curve corresponding to this limiting perihelion. Note that the cones may become quite wide for
the lower velocities.

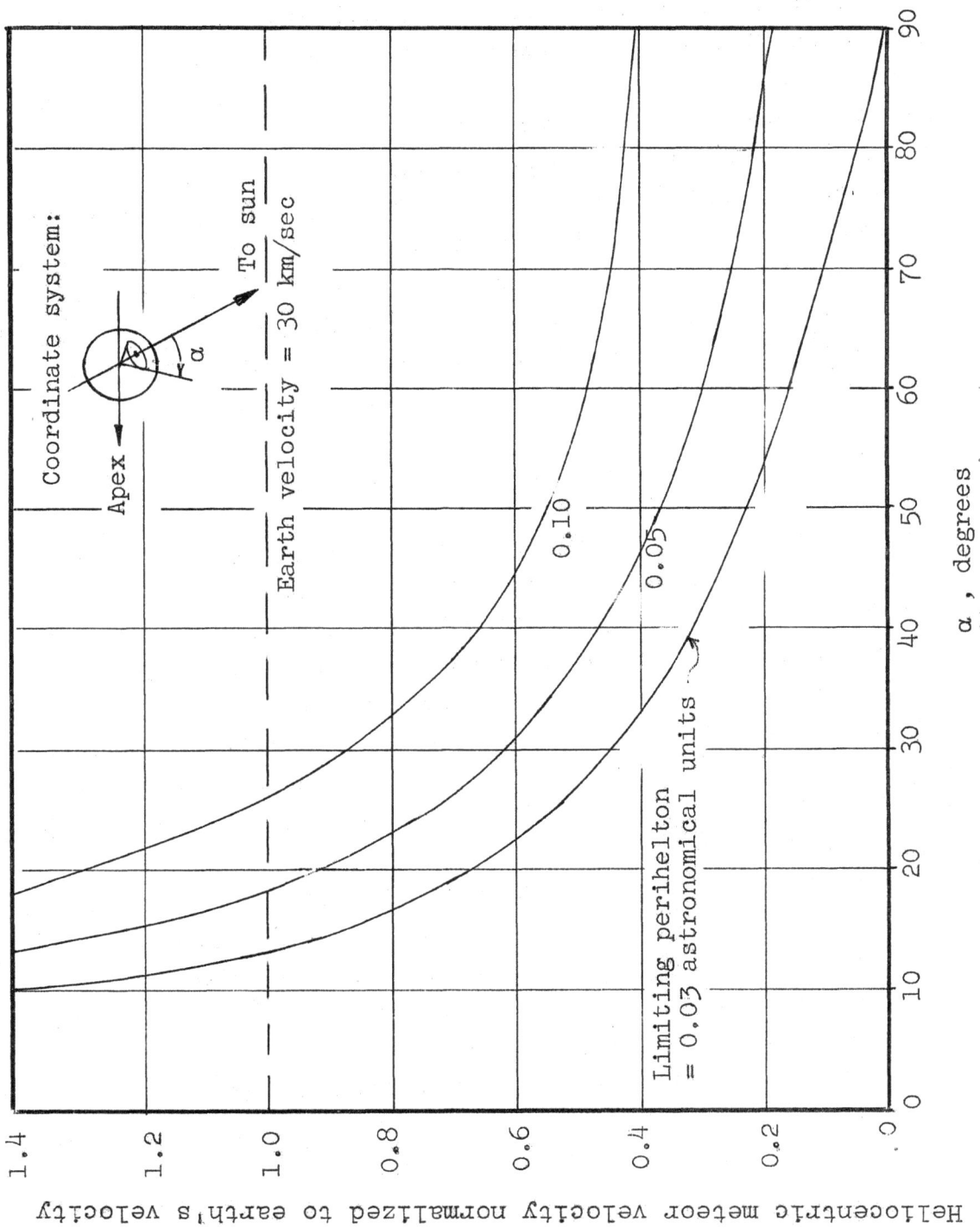

FIG. 12.--Possible velocities of meteors intercepting earth at conical angles α, between earth-sun axis and radiant direction. All velocities above and to the right of each limiting perihelion curve are possible.

When the velocity of the earth is added into the problem
of the dead zones, the cones from which no meteors should emanate
are shifted toward the apex direction. This effect is illustrated in Fig. 13. Considerations of these dead spots are
important in applications to meteor propagation. While a
meteor shower provides an active spot in the radiant distribution,
the zones considered here are devoid of meteors.

Future refinements in theoretical models of the distribution
of meteor orbits should include some degree of concentration of
the orbits into the ecliptic plane, and a consideration of the
dead zones described above. However, before much new information
can be obtained from such investigations, more experimental
data on meteor rates and radiants are needed.

B. ECHO RATES

The radiant distributions and meteor rates considered
above apply to the total celestial hemisphere of the observer.
When meteor trails are detected by radio techniques, by far
the strongest echoes are obtained from the trails which are so
positioned as to provide a specular reflection of the radio
energy from transmitter to receiver. Thus echoes are obtained
from only a small percentage of the trails which are produced
in the common volume in the antenna beams of the transmitter
and receiver.

The radiant selection characteristics of meteor detection
by radio has been treated extensively in the literature. The
point we wish to stress here is the fact that there is a band
or strip of radiants on the celestial hemisphere from which
trails emanate that are potentially detectable at the intersection of single rays from the transmitter and receiver. This
band (and the propagation path geometry) is illustrated in
Fig. 3 of reference 3.

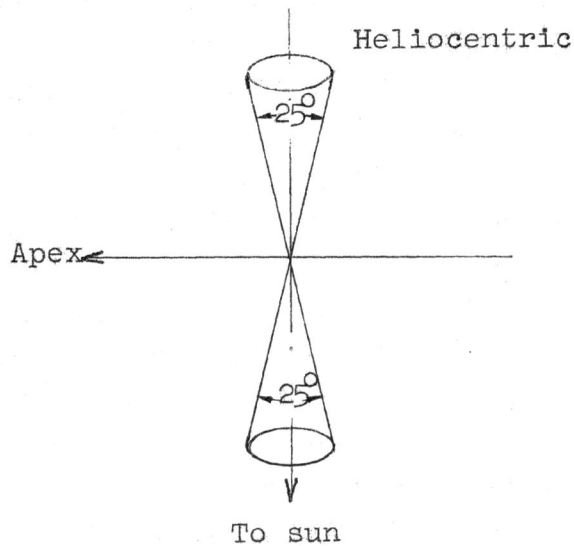

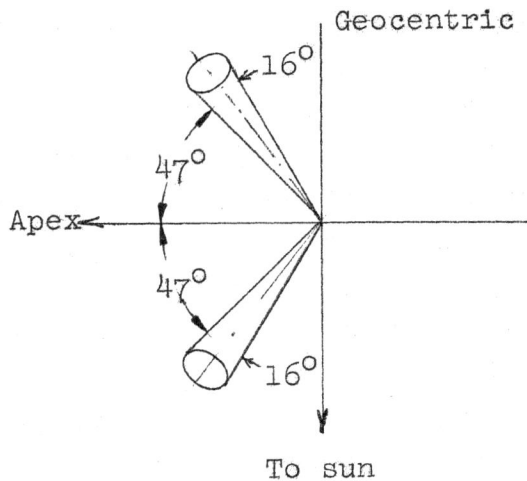

FIG. 13.--Illustration of the conical "dead" zones
in which no meteor radiants lie; heliocentric and geocentric references. It is assumed that the limiting
perihelion for meteors is 0.05 astronomical units
(5 sun diameters), and that the heliocentric meteor
velocities are all 42 km/sec.

The importance of this band-concept in the program of predicting oblique-path meteor propagation from radar measurements can be seen from the following considerations. For a single ray from a radar site, the position of the band of echo-producing radiants is determined by the intersection of a plane perpendicular to this ray with the celestial hemisphere. But meteors from this same band of radiants will produce echoes over the particular oblique paths for which the bisector of the forward scattering angle (the internal angle formed by the rays R_1 and R_2 from transmitter and receiver to the meteor plane) lies in the same direction as the radar ray. The activity of all possible bands on the celestial hemisphere can be sampled by noting the number of radar echoes obtained from all azimuth and elevation angles. The measured activity in these bands can then be used to compute the number of echoes that should be obtained over an oblique path for all azimuth and elevation angles of R_1 and R_2. This technique forms the basis for conducting the radar measurements on this contract. While the direct application of the radar results to oblique meter propagation is of primary concern, it is important to note that the two-dimensional radiant distribution can be determined from complete information on the number of single-station radar echoes obtained from each azimuth and elevation angle. That is, although the radiants of the individual trails are not determined, the radiant distribution can be built up from the radar statistics. (For a more detailed description of the above factors, the reader is referred to references 3, 18, and 19. The publication referred to in reference 18 was published as Scientific Report No. 3.)

While the principles involved in relating radar measurements to meteor propagation are fairly straightforward, some complicating factors are encountered. Because of the screening of half of the celestial sphere by the earth, the activity measured by the radar applies to great-circle bands of $180°$ length. In applying these results to an oblique meteor propagation path, the horizon at the midpoint of the path should correspond to the radar horizon. This error caused by differences in horizons should not be an important problem unless the radar and meteor burst path are at appreciably different latitudes (say more than 10° apart). Longitude differences in the radar and oblique path cause little difficulty since the changing radiant activity, on the average, is based on local time. (The difficulty mentioned above in predicting for a different latitude is based upon the concept of using individual band measurements for the oblique-path prediction. If the total two-dimensional radiant distribution were determined from computations based on many band measurements, then reliable predictions could be made for all latitudes.)

The width of the radiant band for a radar ray is approximately L/R radians, where L is the trail length and R is the radar range. The width of the radiant band for oblique-path propagation is approximately $L(R_1 + R_2) (1 - \sin^2 \phi \cos^2 \beta)$ /2 $R_1 R_2 \cos \phi$ radians. That is, the band for a given trail length and path geometry changes in width with β, the angle between the meteor trails axes and the plane of propagation. Since the width of a radar band does not vary with position on the celestial hemisphere, it might at first appear that an important error is introduced in changing from radar measurement to oblique path prediction. However, it turns out that where the band is narrowest for oblique propagation, the echo intensities are greatest, and where the band is broadest, the echo intensities are the smallest. These factors change in such a way that the product of width (proportional to the number of echoes) and echo intensities is constant for all positions along the oblique path band, just as it is for the radar band. This result can be seen mathematically by multiplying the expressions given above for the widths with the expression for long-wavelength, low-density echo intensity given in Table I, noting the changes given in section IIB for applying the intensity formula to oblique paths.

To demonstrate the manner in which the rate of detection of meteor echoes varies with the direction of a ray from a radar site, computations were carried out for a theoretical radiant distribution of the type used in computing the curves for Figs. 8 and 9. The results of this computation were presented at a scientific meeting (U.R.S.I.) in Washington, D. C., in 1955, and were also presented in Fig. 7 of reference 19. These curves are presented again in Fig. 14 of this report for further consideration and discussion. Figure 14 presents the theoretical diurnal radar rate of meteor echoes for four pencil antenna beams elevated at 45° and directed north, south, east, and west. From the previous discussion, it is apparent that the curves of Fig. 14 should also show qualitatively the number and directional characteristics of echoes observed over oblique paths. In particular, the curves labeled north and south show the relative numbers of echoes arriving at the receiver of an east-west propagation path from directions which are north and south, respectively, of the great-circle path to the transmitter. Similarly, the curves labeled east and west show the relative numbers of echoes arriving at the receiver of a north-south path from directions which are east and west of the great circle path.

The same theoretical results presented in Fig. 14 are presented in a different way in Fig. 15. To illustrate the off-path characteristics of oblique-path meteor propagation over east-west and north-south paths, the ratios $(N-S)/(N + S)$ and $(E-W)/(E + W)$ are determined from the N, S, E, and W curves in Fig. 14. These ratios are plotted in Fig. 15. For times of day when $(N - S)/(N + S)$ is positive, more echoes can be obtained over an east-west path by pointing the transmitting and receiving antennas to the north side of the path, than by pointing them south of the path. The reverse is true when this ratio is negative. Similar conditions apply for a north- south path and the $(E - W)/(E + W)$ ratios. It is seen that the predicted optimum off-path direction for an east-west path is north of the great-circle path in the morning, and south of the

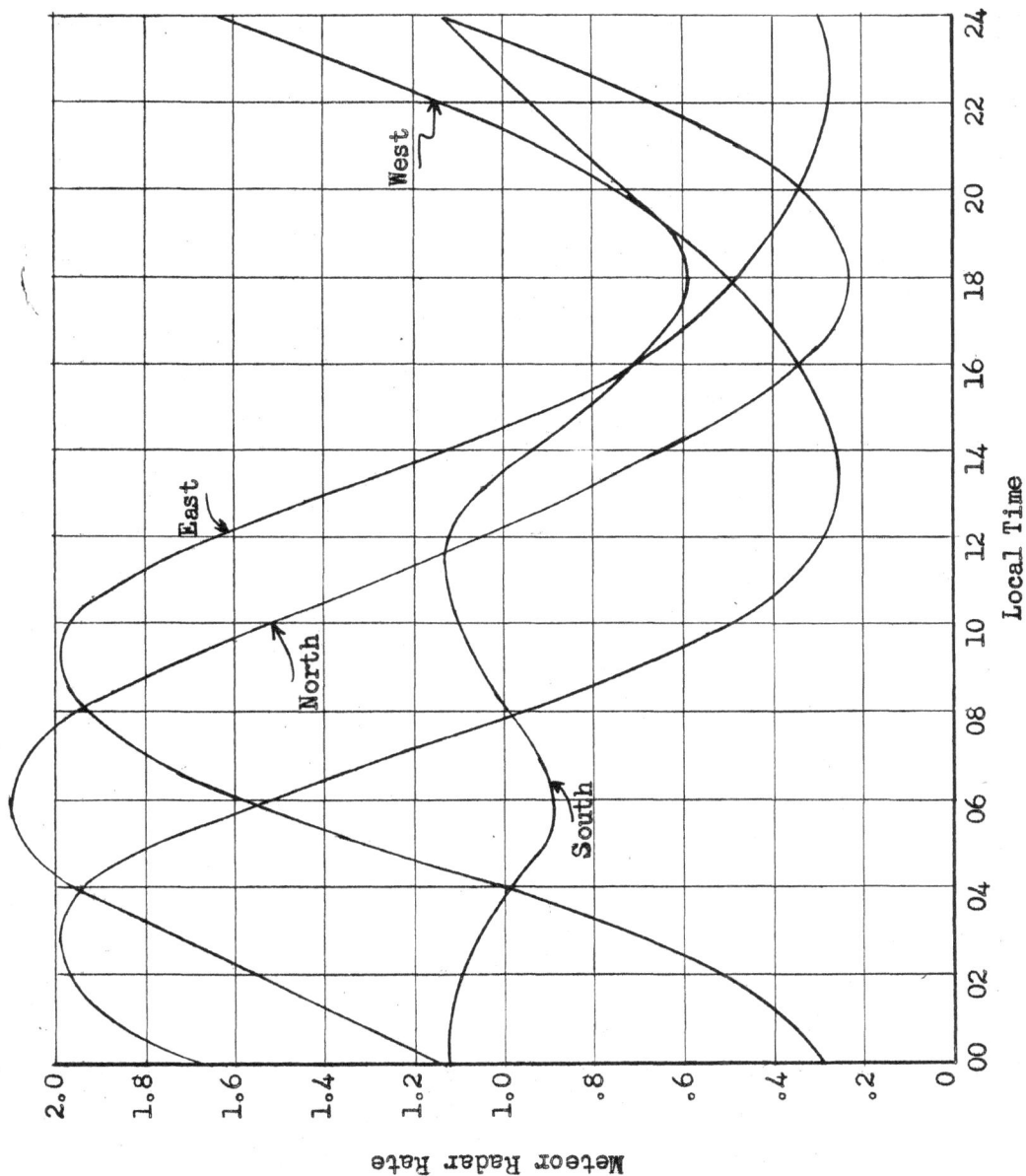

FIG. 14.--Theoretical diurnal radar rate of meteor echoes for a pencil antenna beam elevated at 45° and directed north, south, east, and west. A spherically-uniform heliocentric distribution of parabolic meteor orbits is assumed, and the results are for 37.5° N latitude at the summer solstice.

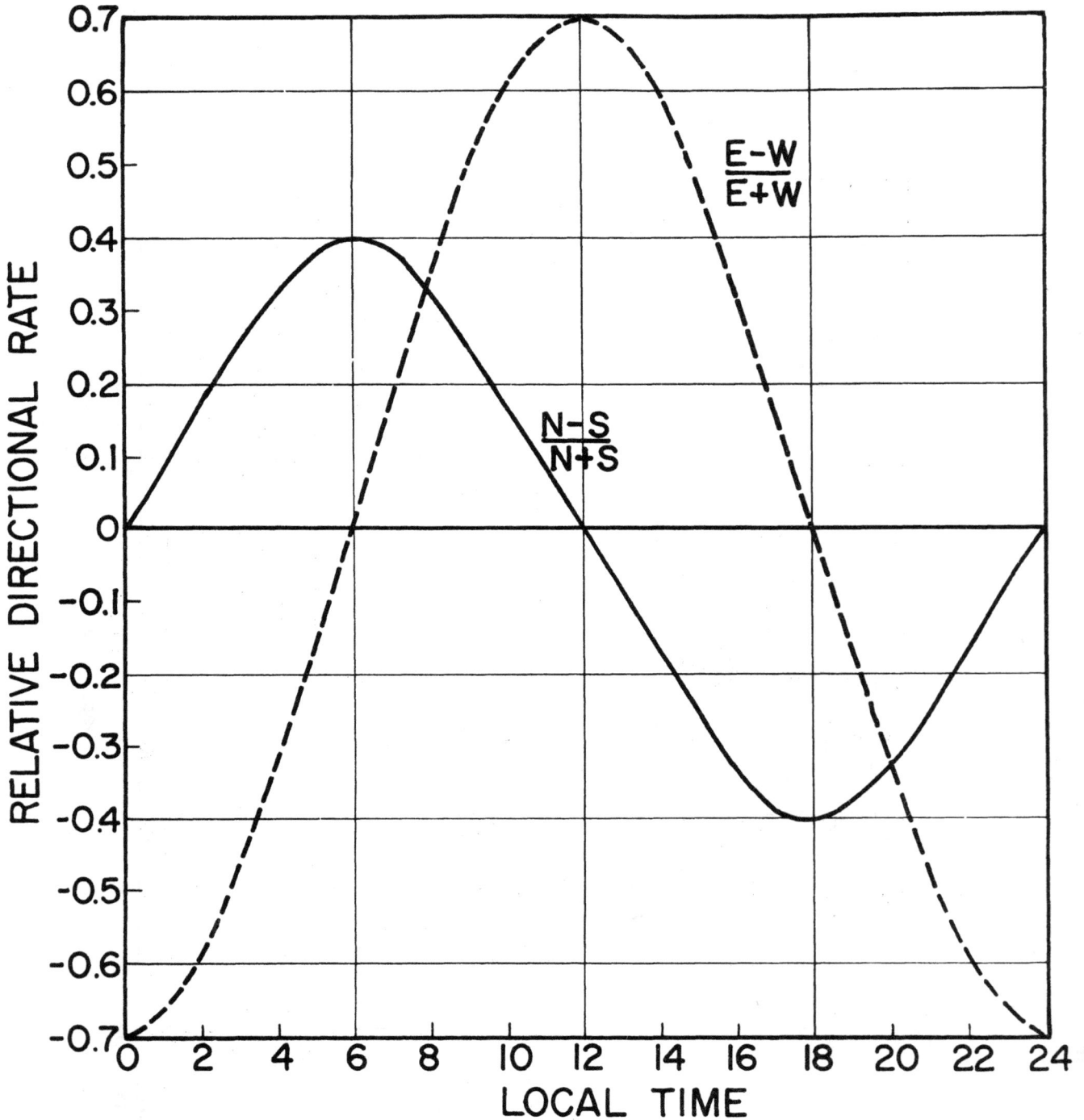

FIG. 15.--Theoretical (N-S)/(N+S) and (E-W)/(E+W) ratios for a radar at northern temperature latitudes. N, S, E, and W refer to the total number of echoes obtained with a pencil radar beam pointed north, south, east and west respectively at an elevation angle of 45°.

path in the evening. It is west of the great-circle direction of a north-south path at night, and east of the path in the day. The amount by which the ratios go positive or negative is a measure of the relative importance of using the optimum side of the path.

The above predicted gross characteristics of the directional properties of meteor-burst propagation are presented in a very simple manner in Figs. 16 and 17.

It should be noted that for stations in the southern hemisphere, the above results apply when the north and south labels are interchanged.

The directional characteristics of meteor propagation are very important in both meteor-burst and ionospheric-scatter propagation. The theoretical predictions discussed above are compared with measurements in the next section.

V. RADAR MEASUREMENTS OF METEOR IONIZATION TRAILS

A 50-kw peak-power radar system operating at a frequency of 38 Mc was designed, built, and operated under this project. With this system the range and azimuth of meteor echoes are determined using a directional antenna which is programmed to sample all directions every half hour. The elevation angle of the echoes can be determined roughly from the measured slant range, since it is known that meteors occur in a fairly limited height range. It was pointed out in the preceding section that radar measurements of azimuth and elevation angles of meteor echoes can be used to predict characteristics of meteor propagation over oblique paths. The definition possible in the radar measurements of the radiant distribution is limited by the antenna beamwidth (approximately 20°), the spread in meteor heights, and the lengths of the meteor ionization trails.

E-W Path, Northern Hemisphere

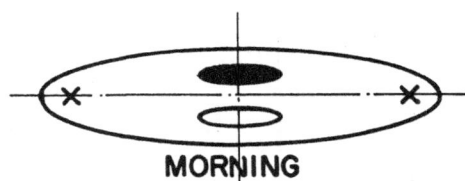

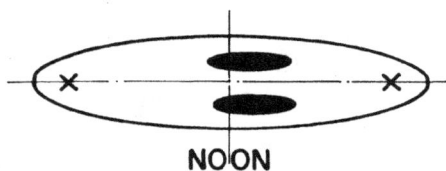

MIDNIGHT

MORNING

NOON

EVENING

Plan View of Favorable Meteor-Scatter Area

FIG. 16.--Spherically-uniform heliocentric distribution of
parabolic meteor orbits.

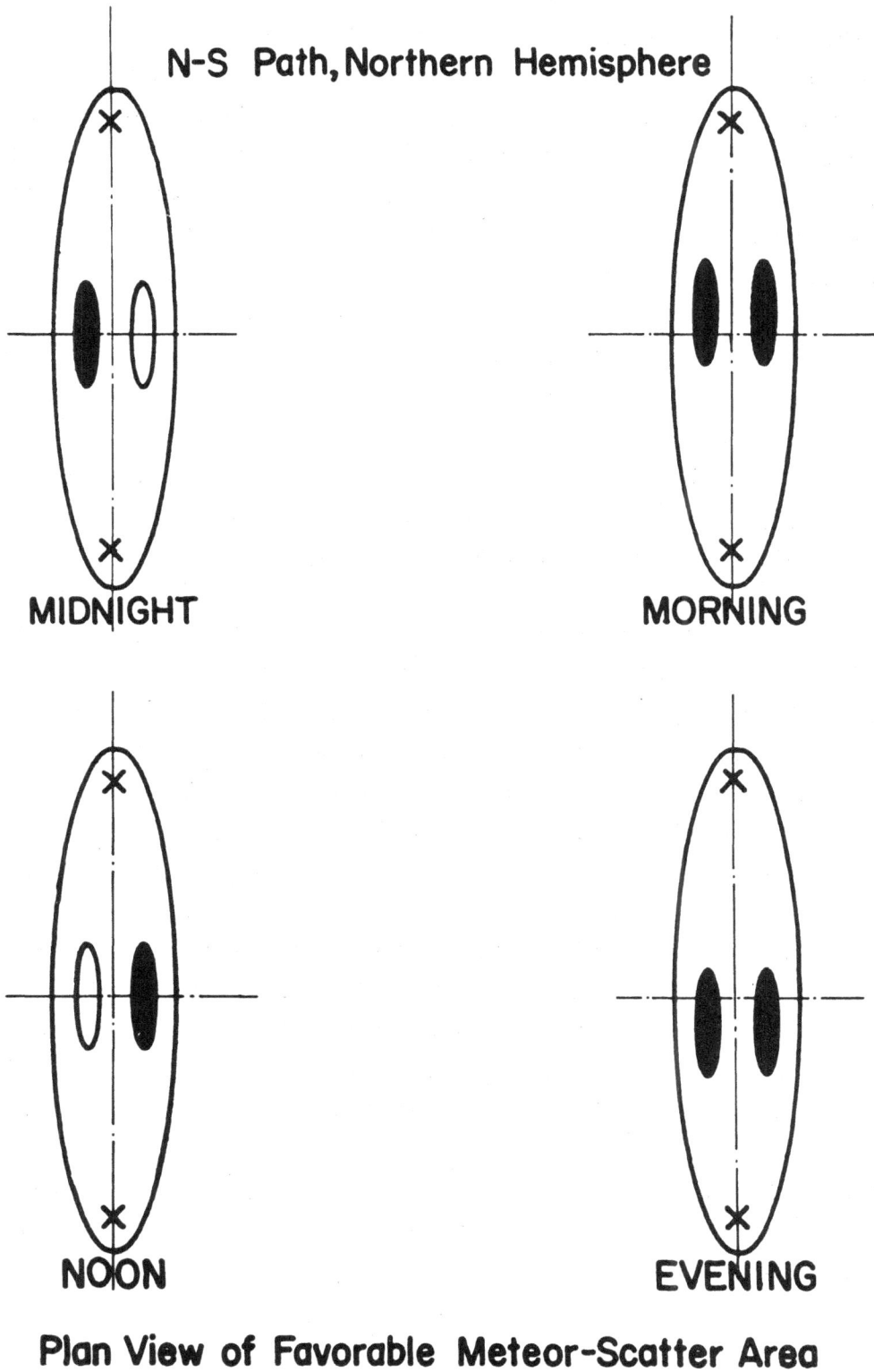

N-S Path, Northern Hemisphere

MIDNIGHT

MORNING

NOON

EVENING

Plan View of Favorable Meteor-Scatter Area

FIG. 17.--Spherically-uniform heliocentric distribution of parabolic meteor orbits.

The first antenna system used with the 38 Mc radar con-
sisted of a nest of six corner reflectors, individually fed by
quarter-wave unipoles over ground planes. This antenna is
pictured in Fig. 18. The various corners were excited for
10 minutes each hour from the transmitter through a coaxial
switch. The same antenna was used for transmitting and
receiving. Considerable delay was encountered in bringing this
antenna into continued operation because the coaxial switch
developed and built at the university failed after a few weeks
of operation. A commercial coaxial switch was then ordered,
but nearly 6 months elapsed before delivery. About a month
after it was put into operation, the wind and rain storms of
December, 1955 destroyed the corner reflectors. It was also
found that the commercial coaxial switch was no longer operable
at this time.

A new rotatable antenna system was then designed, and bids
were let for its construction. The completed array, pictured
in Fig. 19, was delivered in May, 1956. The final installation
of this antenna and its associated control circuitry was completed in August, 1956. The radar system has been in relatively
continuous operation since that time.

For the reasons outlined above, the radar measurements of
meteor azimuth and range distributions have been obtained
discontinuously during several different periods of time since
January, 1955. The data are not sufficient to establish
firmly the seasonal variations of meteoric activity. However,
many measures of the diurnal variation of meteoric activity
have been obtained. The first measure of the diurnal variation
in the azimuthal direction of arrival of meteor echoes was
obtained with this equipment on January 26-27, 1955. The
results of this test were compared with theoretical predictions
in Fig. 5 of Scientific Report No. 3.

FIG. 18.--Nest of 6 corner reflector antennas used with the rate and radiant radar. These antennas were destroyed by wind and rain storms in December, 1955.

FIG. 19.--Rotating antenna used with the rate and radiant radar since August, 1956.

To illustrate the type of results obtained and the analyses used in the radar program, Figs. 20 through 31 have been prepared. In Fig. 20 is presented the total echo rate as a function of time for 48 hours on September 1-3, 1956. The same measurements are presented in Figs. 21 and 22 to show the diurnal changes in the number-azimuth distribution of the echoes, and in Figs. 23 and 24 to show the diurnal changes in the number-range distribution. In Figs. 25 and 26, the diurnal echo rate for the same days is shown for all azimuthal directions which are north, south, east, and west of the radar site, respectively. The ratios $(N - S)/N + S$ and $(E - W)/(E + W)$, whose application to oblique-path propagation was discussed in the previous section, are formed from the radar measurements and presented in Fig. 27. We note that the general trends predicted in the theoretical developments in the preceding section are observed in the radar results. Figure 27, in particular, reveals that the side of an oblique propagation path from which most meteor echoes can be obtained changes diurnally in a manner similar to the predicted variation.

While the gross features of the observed diurnal variations of the meteoric activity are stable, consideration of the data obtained on many days of operation shows that there are important, short-time variations in the number and directional statistics of the echoes. For application of the radar results to oblique-path meteor propagation, the first-order predictions for the optimum azimuths of the antenna beams are now available from the gross radar characteristics. However, if it is desired to use the apparent short-term variations in the radiant distribution, it appears that a technique of continuous monitoring of the meteoric activity would be required. For instance, by providing a radar system at one terminal   a meteor-ionization propagation path, the changing meteoric activity could be measured, and the results of these measurements could be used immediately to determine the optimum bearing for the antennas used in the oblique path. Or, the changing meteoric activity could be monitored over the oblique path itself by use of a separate transmitting and receiving system.

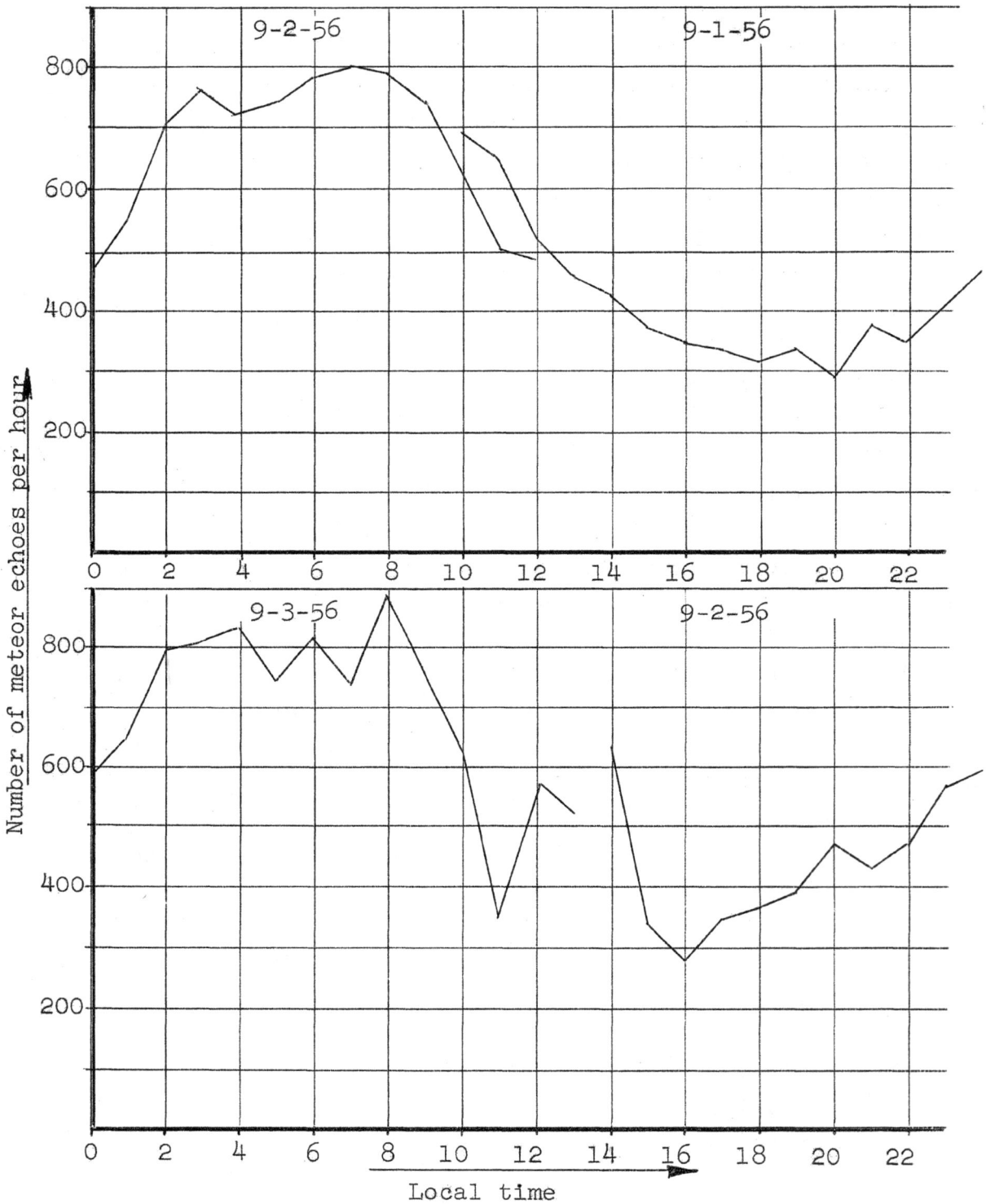

FIG. 20.--Diurnal variation of total meter
echo rate; Sept. 1-3, 1956.

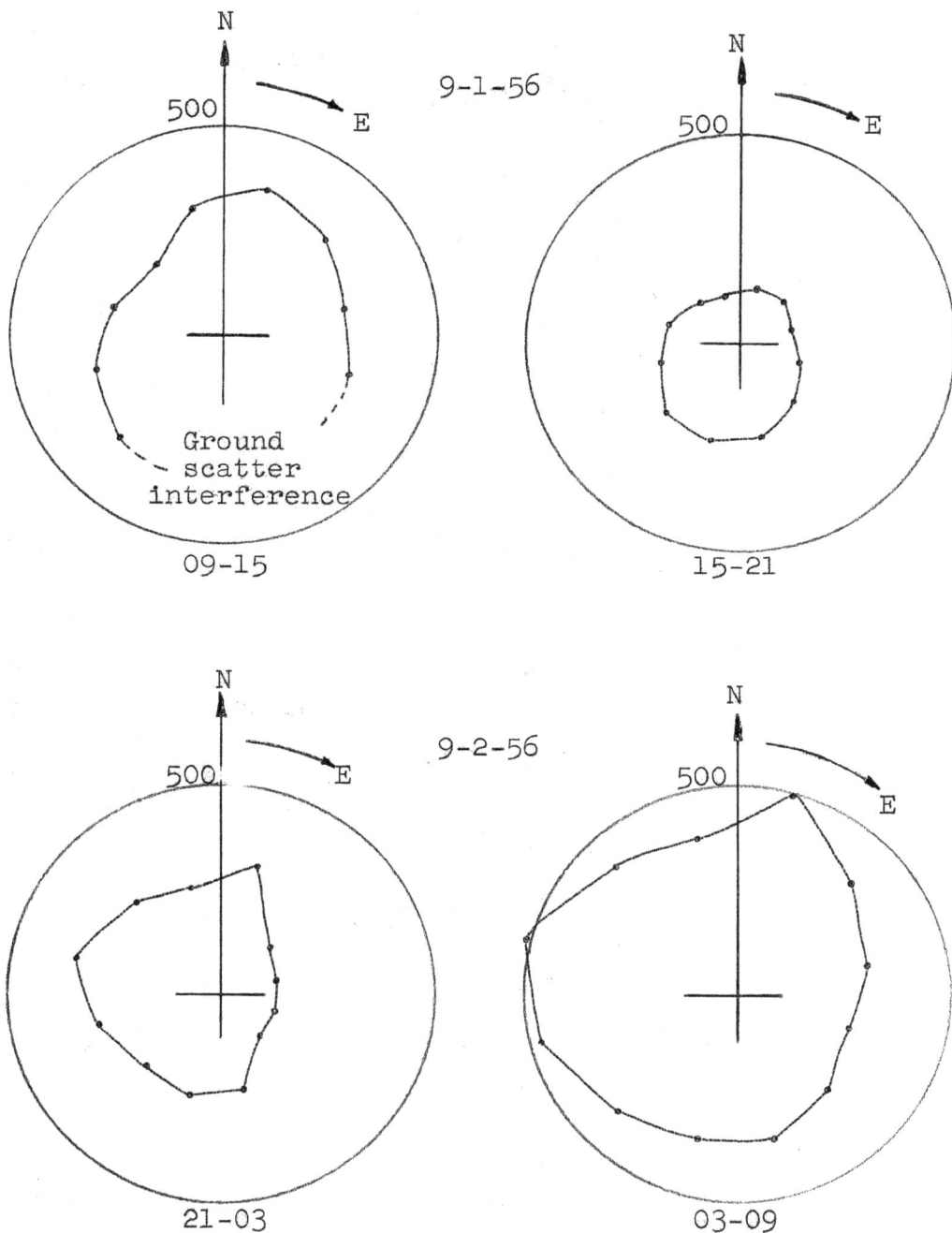

FIG. 21.--Azimuthal distribution of meteor echo rate
measured on Sept. 1-2, 1956; six hour periods. The
circle radius corresponds to an echo rate of 500 echoes
in the six hours in a 30° sector.

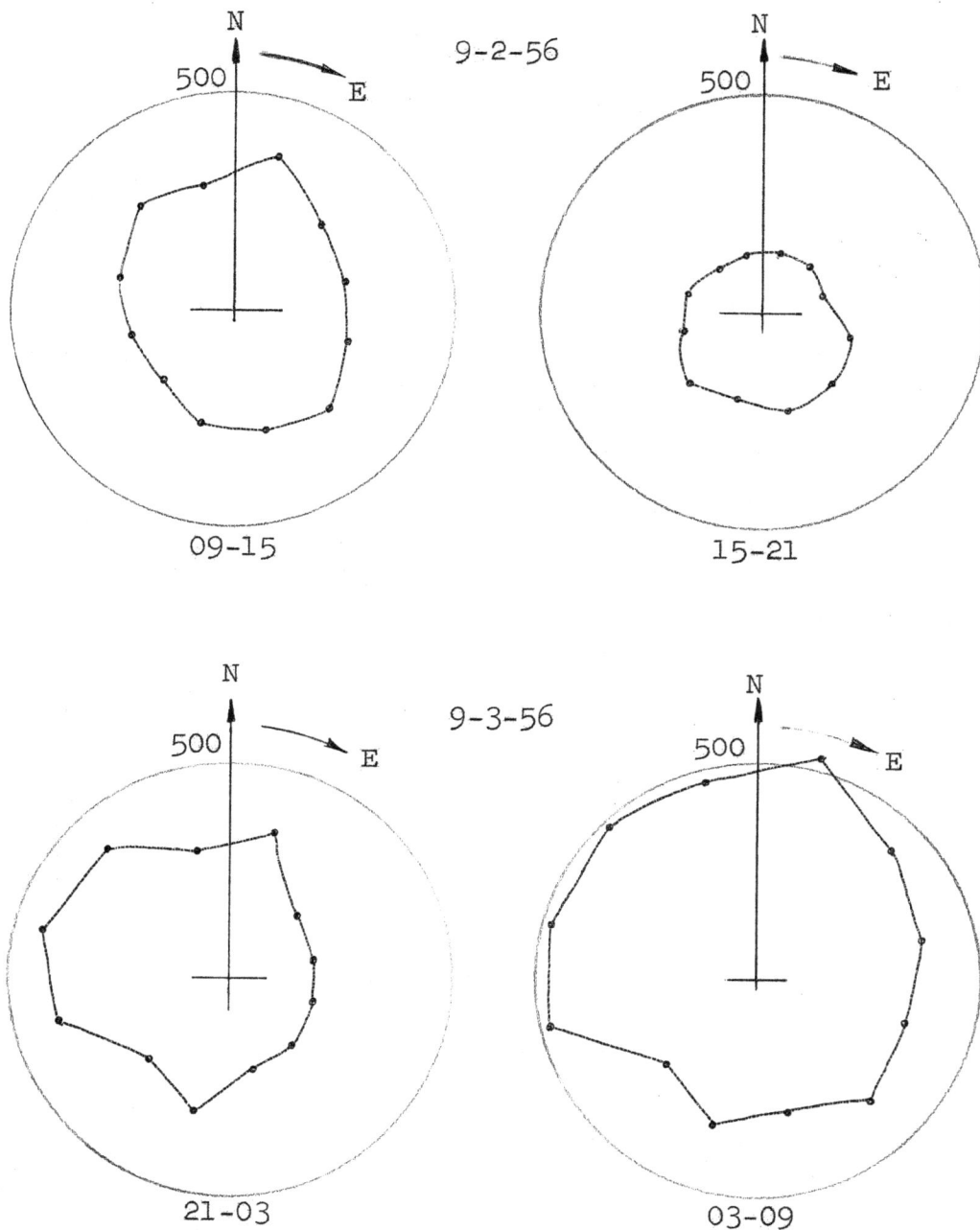

FIG. 22.--Azimuthal distribution of meteor echo rate
measured on Sept. 2-3, 1956; six-hour periods. The
circle radius corresponds to an echo rate of 500
echoes in the six hours in a 30° sector.

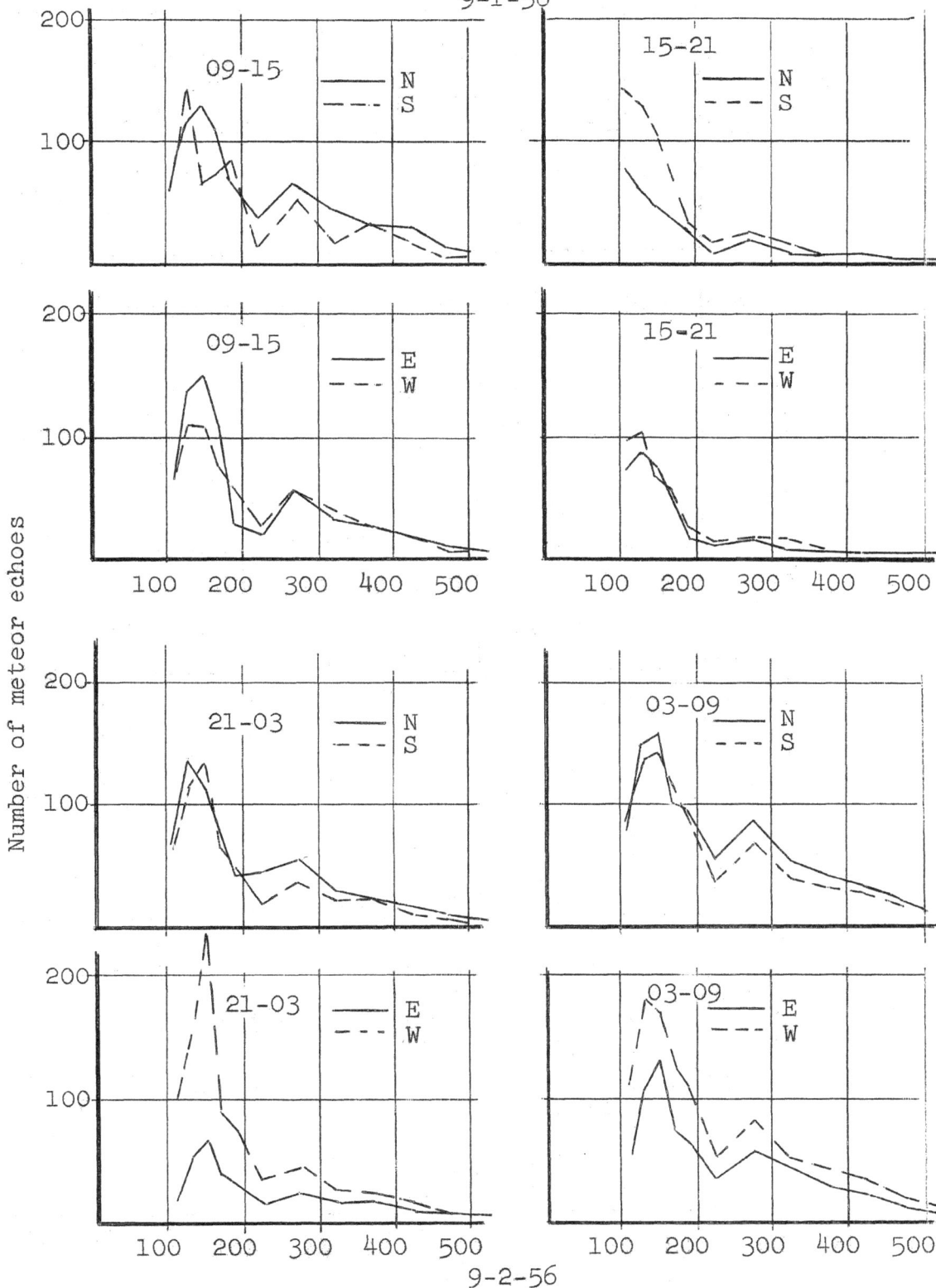

FIG. 23.--Range distribution of the meteor echoes in the
N, S, E, and W quadrants for six-hour intervals; Sept, 1-2,
1956. The indicated rates are for 20 km range intervals.

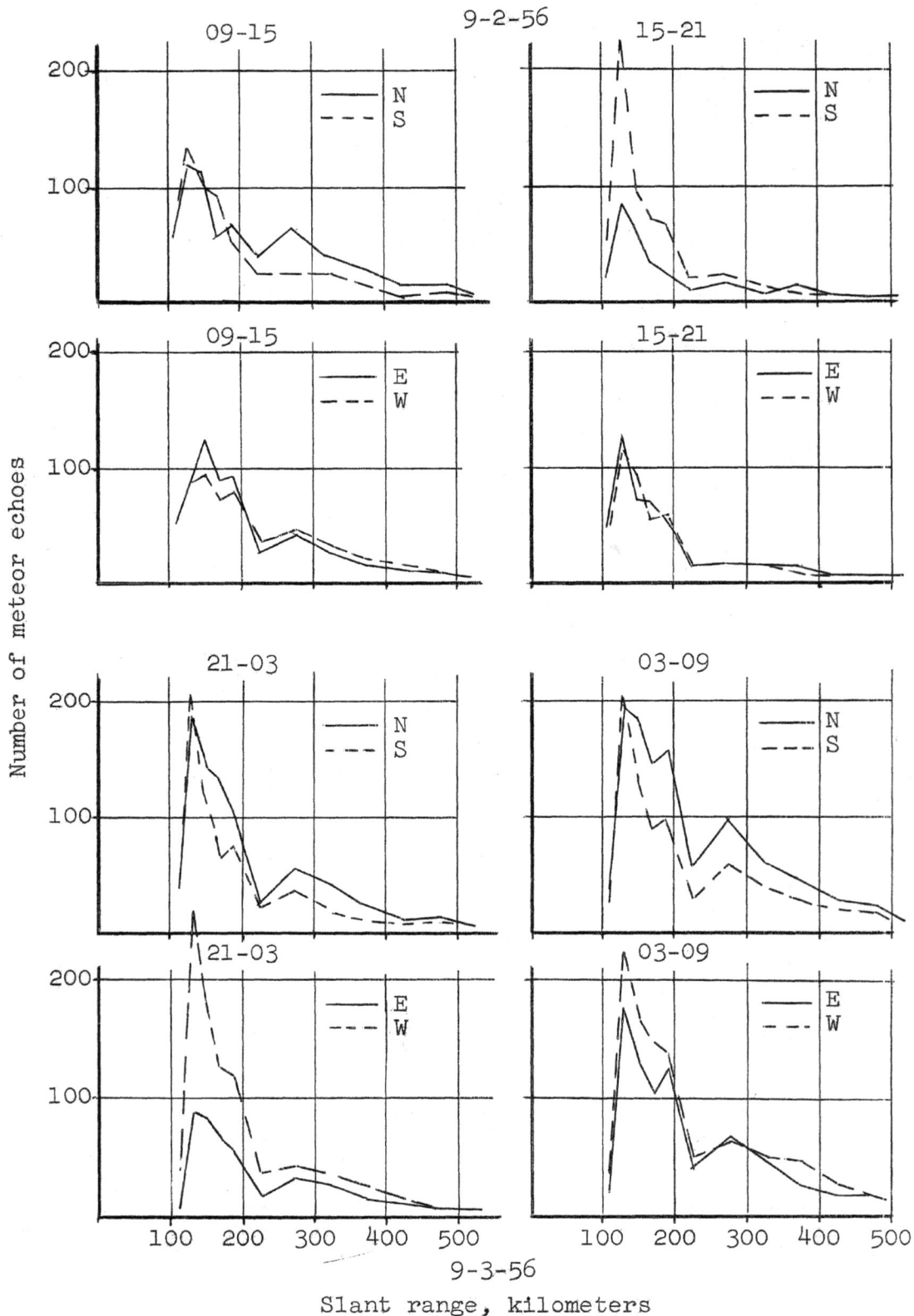

FIG. 24.--Range distribution of the meteor echoes in the
N, S, E, and W quadrants for six-hour intervals; Sept. 2-3,
1956. The indicated rates are for 20 km range intervals.

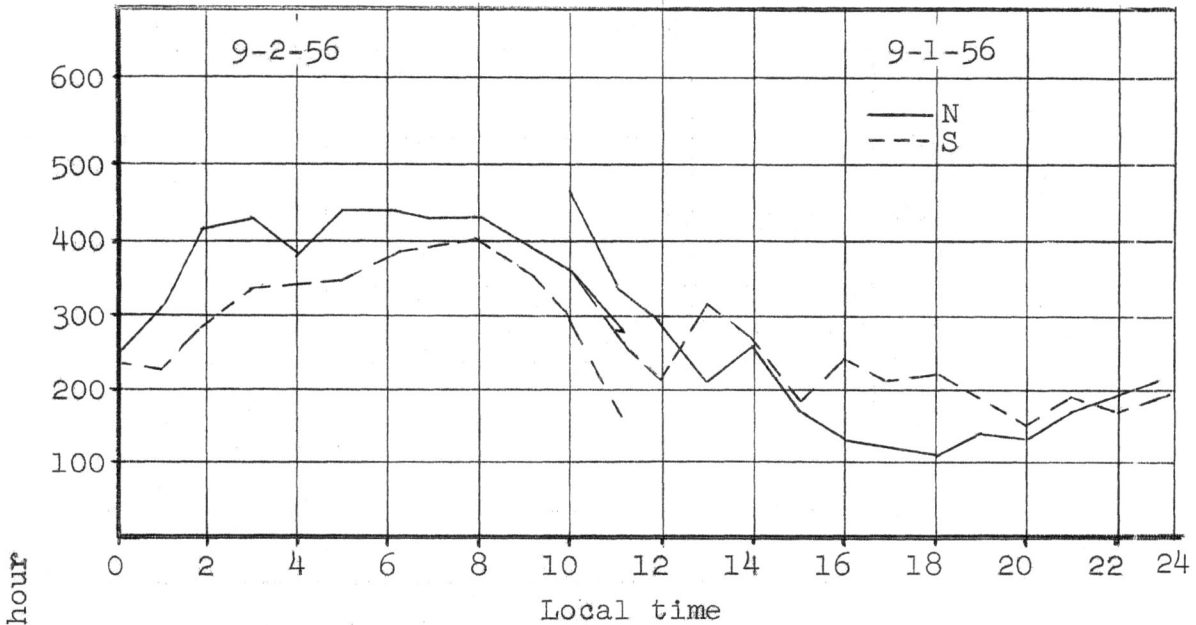

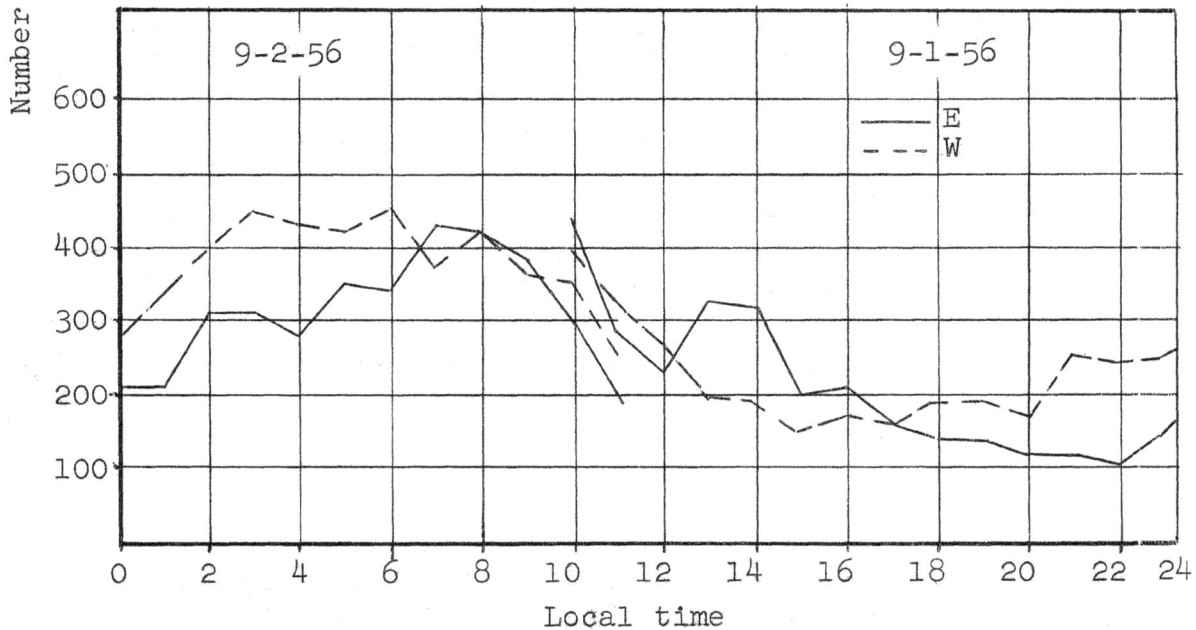

FIG. 25.--Diurnal valuation of meteor echo rate for all
bearings N, S, E, and W of the radar station, respectively; Sept. 1-2, 1956.

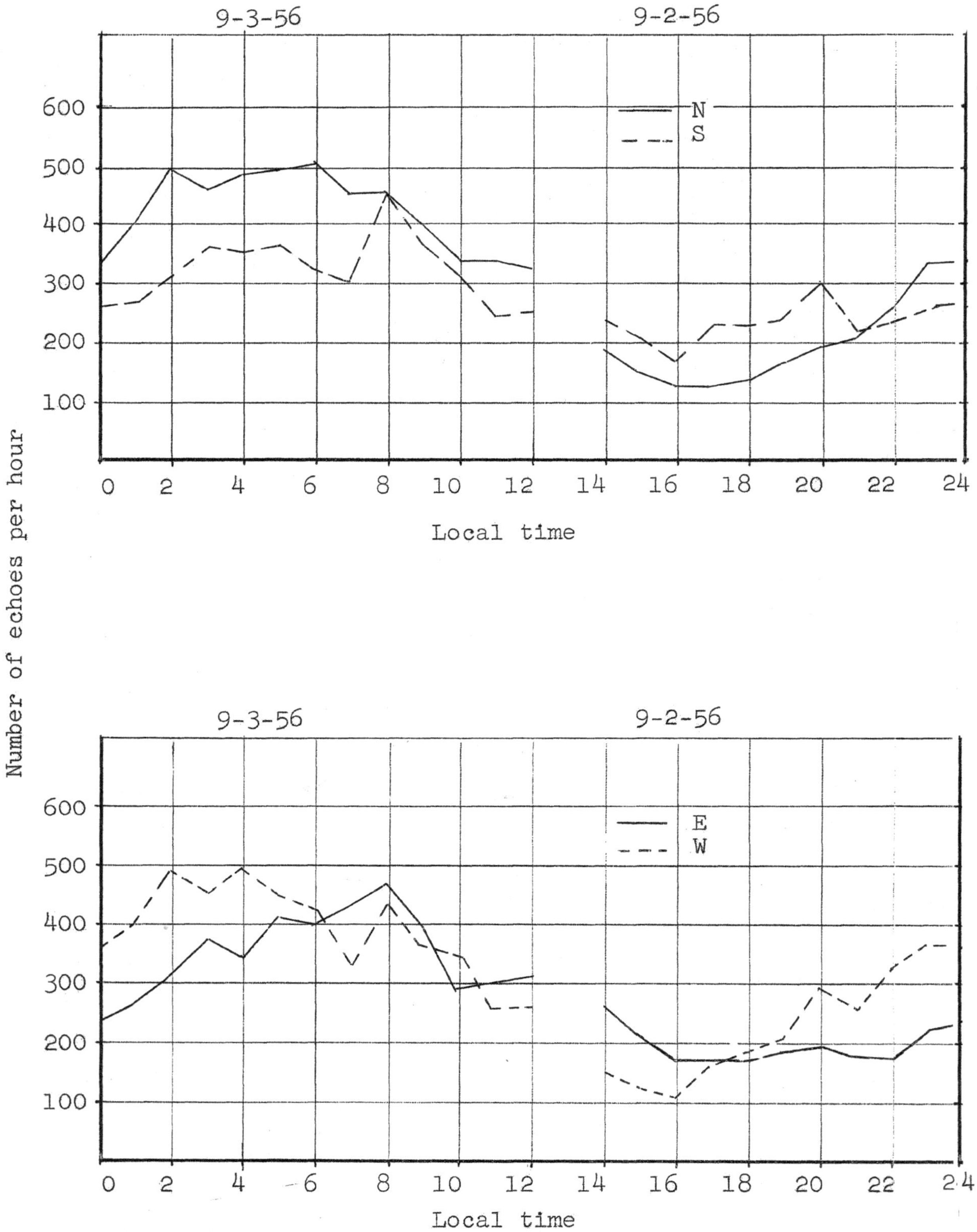

FIG. 26.--Diurnal variation of meteor echo rate for all
bearings N, S, E, and W of the radar station, respectively;
Sept, 2-3, 1956.

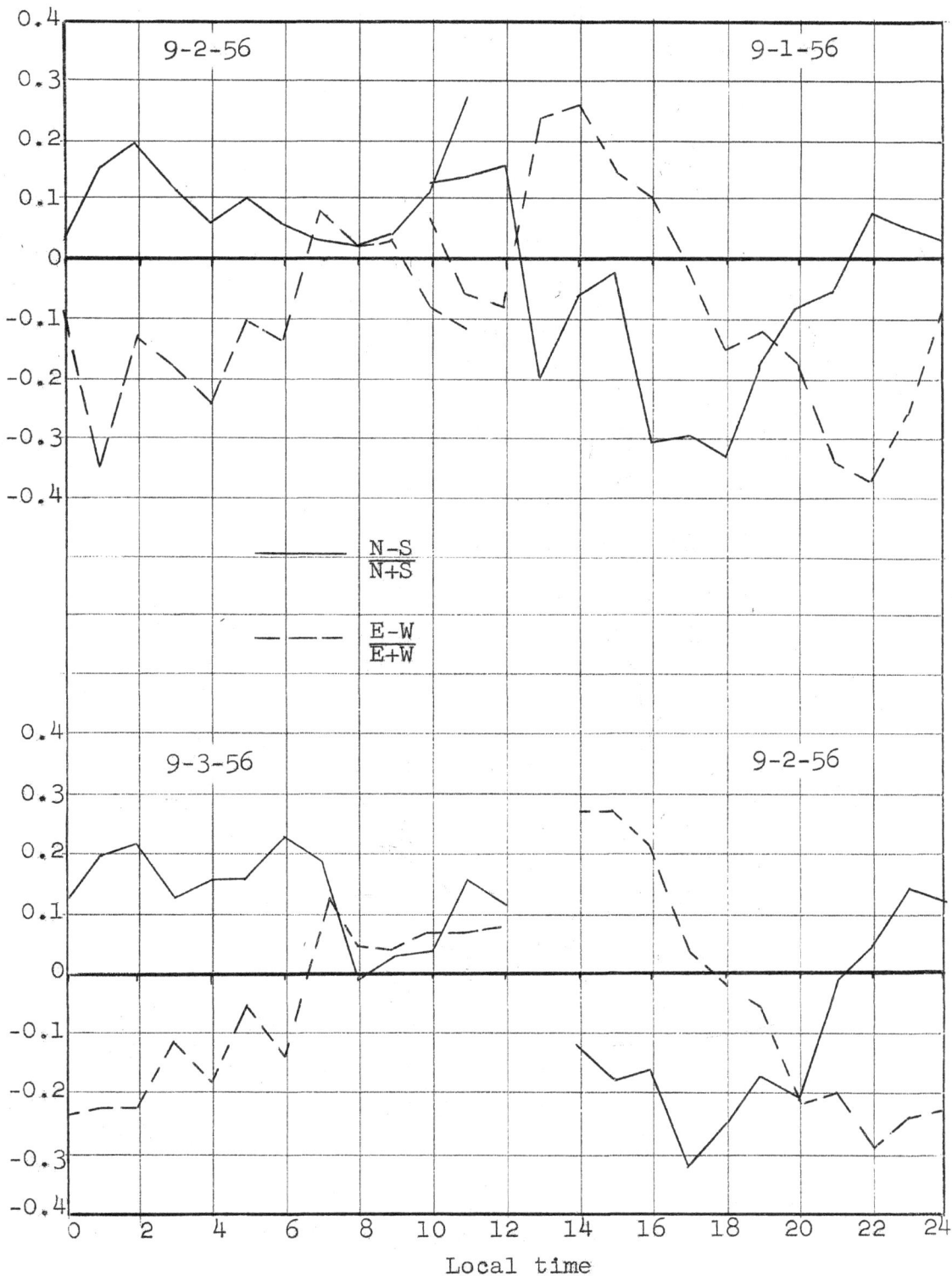

FIG. 27.--Diurnal variations of the ratios (N-S)/(N+S) and
(E-W)/(E+W) for Sept. 1-3, 1956. The values of N, S, E,
and W are the number of echoes from all bearings north,
south, east, and west of the radar station, respectively.

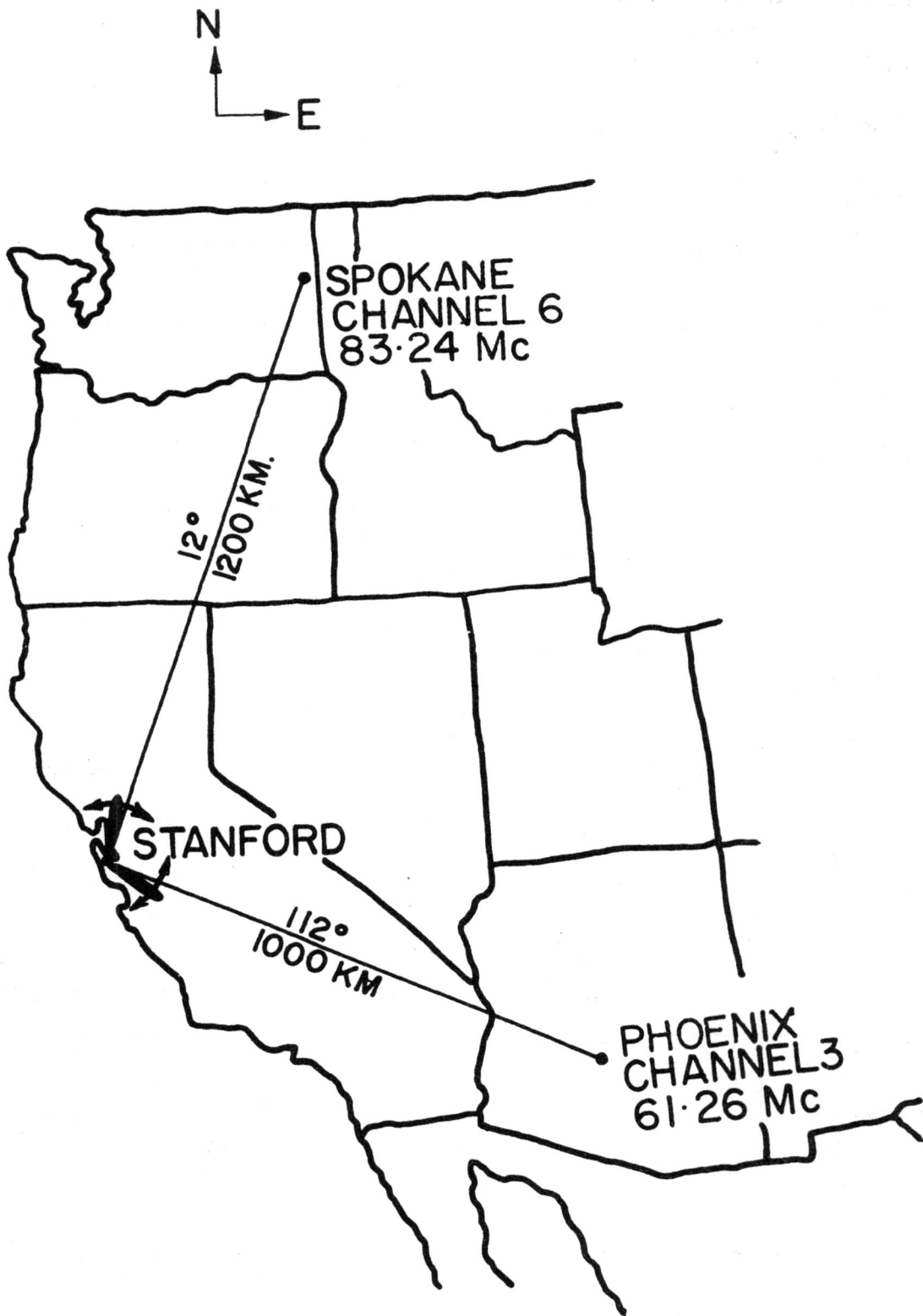

FIG. 28.--Locations of the television stations used in the simultaneous radar and oblique-path measurements of the optimum antenna directions for meteor propagation.

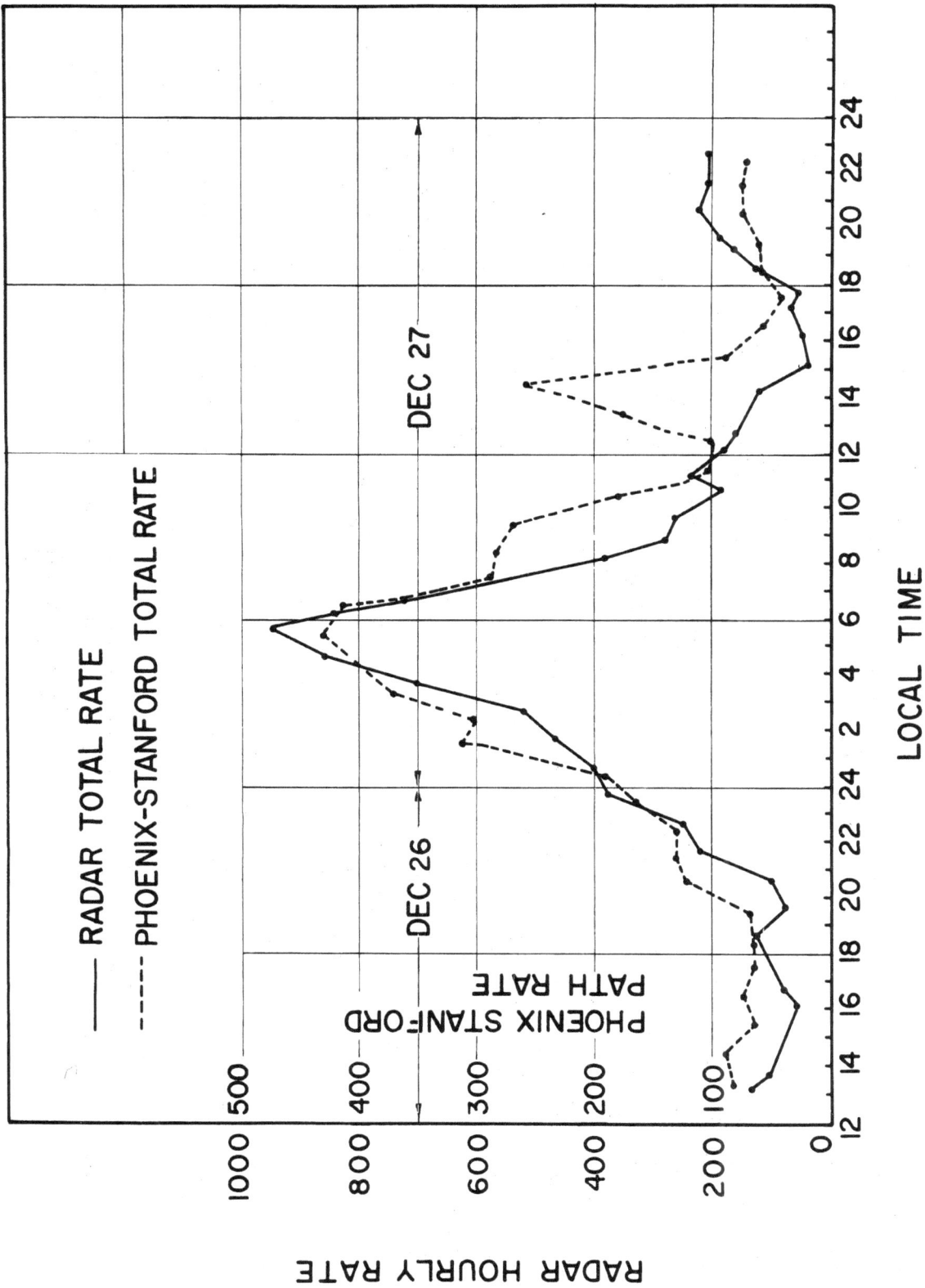

FIG. 29.--Comparison of the diurnal variation of the total meteor echo rate measured with a radar and measured over an east-west oblique path.

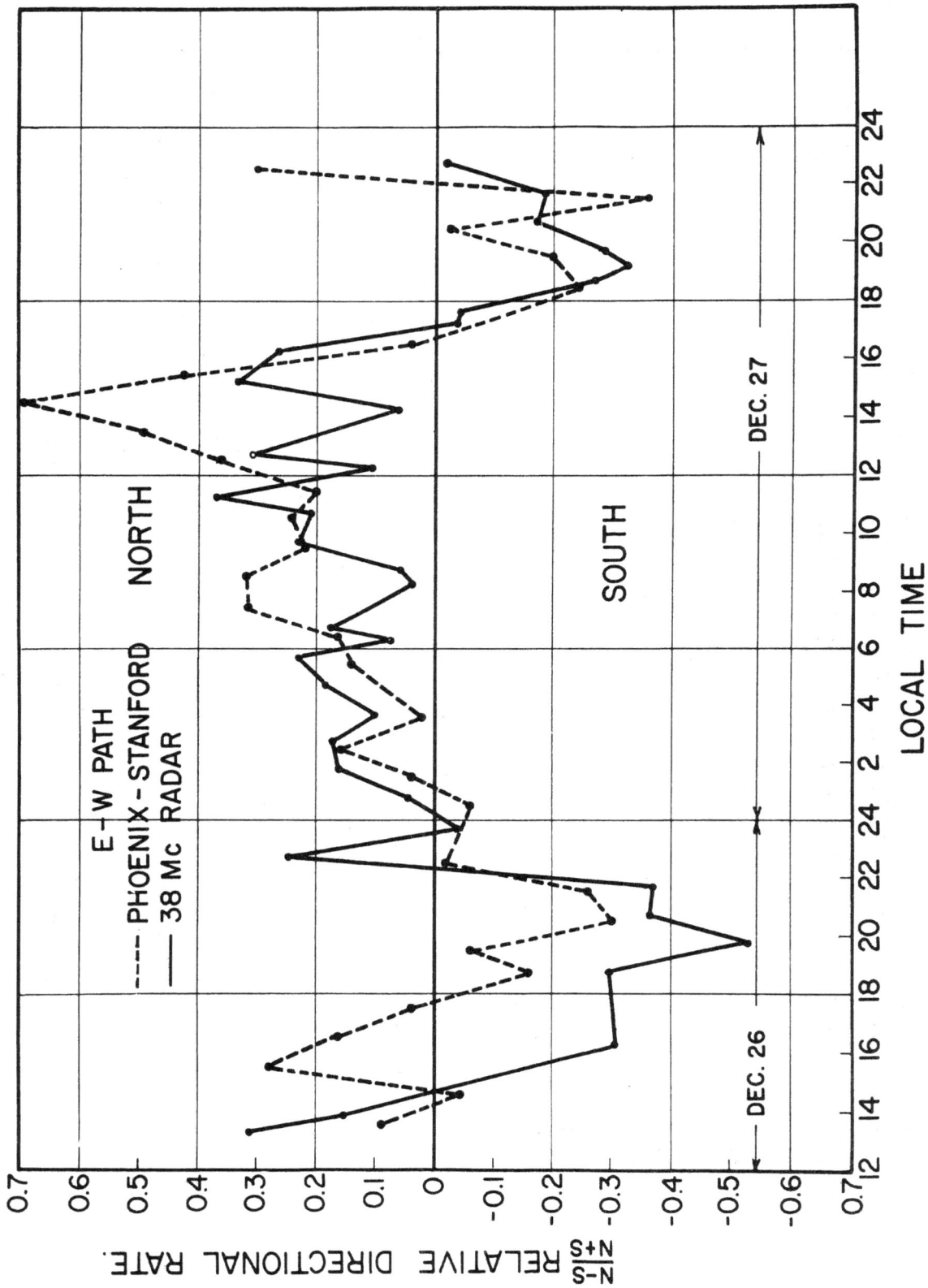

FIG. 30.--Comparison between back scatter and forward scatter measurements of the diurnal variation in the (N-S)/(N+S) ratio. N and S refer to the total number of echoes received from the north and south respectively of the radar station and the north and south sides respectively of the east-west oblique path.

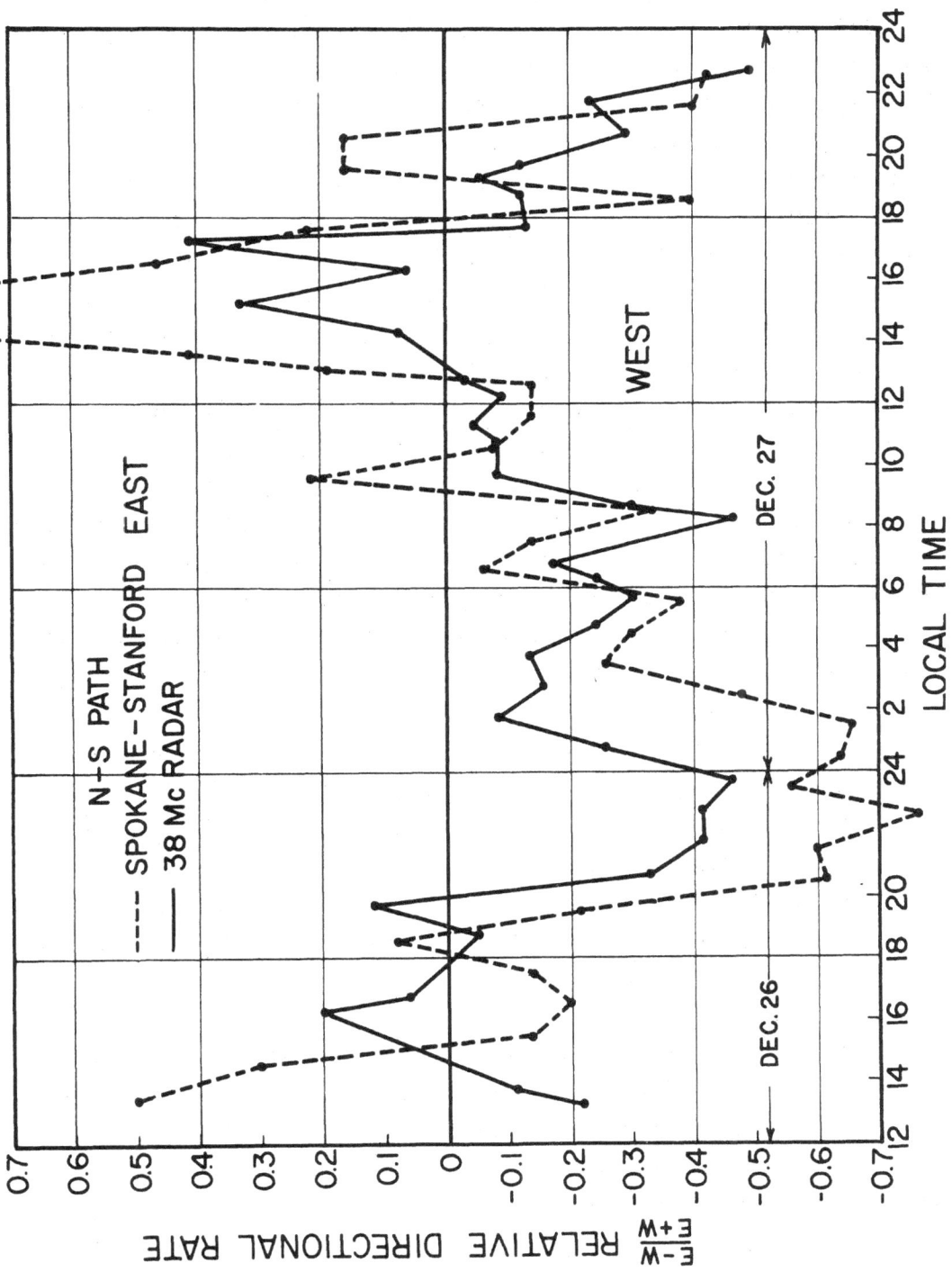

FIG. 31.--Comparison between back scatter and forward scatter measurements of the diurnal variation in the (E-W)/(E+W) ratio. E and W refer to the total number of echoes received from the east and west respectively of the radar station, and the east and west sides respectively of the north-south oblique path.

The feasibility and desirability of an "instantaneous prediction service" as described above has been investigated in a preliminary way in cooperation with the Stanford Research Institute. On December 26 - 27, 1956, the meteor radar was operated at the same time that the directional properties of meteor-burst propagation were being measured by S. R. I. over paths from Phoenix, Arizona, and Spokane, Washington, to Stanford, California. The two oblique paths are pictured in Fig. 28. The Phoenix-Stanford path is approximately E - W, and the Spokane-Stanford path is approximately N - S. The radar information has not been completely analyzed, but for a rough comparison with the oblique-path data, the total number of meteors observed at all bearings north, south, east, and west of the radar have been determined. In Fig. 29, the total meteor rates observed with the radar and over the Phoenix-Stanford path are compared. The magnitudes of the rates are expected to be different, but the diurnal variations should be similar. In Fig. 30, the $(N - S)/(N + S)$ ratios for the radar and for the Phoenix-Stanford path are compared. In Fig. 31, the $(E - W)/(E + W)$ ratios for the radar and for the Spokane-Stanford path are compared. In view of the very rough nature of the preliminary analysis of the radar results, it appears that the agreement between the radar and oblique-path measurements are quite good. More detailed analyses of radar results should make it possible to determine not only the favored side of the path for meteor propagation, but also the exact variation of rate with off-path angle, and with elevation angles at the transmitter and receiver.

There is a very large anomaly in the oblique-path results at about 1400 on December 27. From the data obtained in these measurements made on the Stanford Research Institute project, it was found that the number of echoes arriving from a particular azimuth increased by a factor of ten in less than an hour. After less than two hours, the rate was back to normal. This sudden enhancement, persumably associated with a heretofore undetected meteor shower of very short duration, was not repeated on the preceding or following day. With an "instantaneous prediction" arrangement, it should be possible to make use of such meteoric enhancements in meteor-burst and ionospheric-scatter propagation. Unfortunately, in this example, the radar did not record the increased rate. Computations based on the angle-of-arrival data for the two oblique paths indicate that this shower would produce echoes beyond the maximum range (500 km) at which the radar was recording echoes. More tests of this type should be conducted in future investigations of meteoric phenomena as applied to meteor propagation. It is believed, however, that the feasibility of the radar technique for predicting meteor propagation has been demonstrated.

VI.   PROPERTIES OF METEOR ECHOES PROPAGATED OVER LONG PATHS

Meteor-burst propagation over various paths at different frequencies has been studied by making use of regular and special night-time transmissions from television stations throughout the western half of the United States. Dr. Allen M. Peterson of Stanford University and the Stanford Research Institute demonstrated in 1954 the feasibility and potentialities of this technique for studying meteoric phenomena. Of the receiving equipments used in the tests reported here, some were acquired under this contract and under contract N6onr 251-7D at Stanford University. In most of the tests, however, the equipment developed and built at the Stanford Research Institute under contract AF 19(604)-1517 was used for receiving and recording. The close cooperation of Dr. Peterson and W. R. Vincent of the S. R. I. staff in the tests reported here is gratefully acknowledged.

The special features of meteor propagation which were investigated in the TV recording tests include (1) the dependence of echo duration on path length and radio wavelength, and (2) the number-intensity and number-duration distributions of echoes propagated over a 2000-km path. (The direction-of-arrival tests, mentioned in the previous section in connection with the radar measurements, were conducted entirely by the S. R. I. staff, and are not part of the researches undertaken under this contract. Close liason is maintained with the people planning and conducting these tests, however, because of the very close connection between the directional results and the radar measurements on the meteor radiant distribution.)

A. DEPENDENCE OF ECHO DURATION ON PATH LENGTH AND RADIO WAVE-LENGTH

The locations of the television stations used in the various tests of echo-duration dependence are shown in Fig. 32. In order to minimize the number of nonessential variables in the determination of the duration dependence on path length, four stations operating on the same channel (2) and located approximately east of Stanford were chosen for the main sequence of tests. All of the signals were recorded between the hours of about 0200 and 0400 PST, the transmitters being operated during these normally-off hours by special arrangements with the TV stations. While it would have been desirable to conduct all of the tests on the same night (to avoid possible day-to-day variations), such an arrangement was not feasible with the available receiving equipment and techniques. The tests with the four transmitter locations were conducted during the last two weeks of March, 1956.

The S. R. I. receiving antenna consists of four commercial TV antennas in a broadside array, having a beamwidth of about 20°. This beam was directed toward the station being monitored in each of the tests. The center of the receiving antenna is

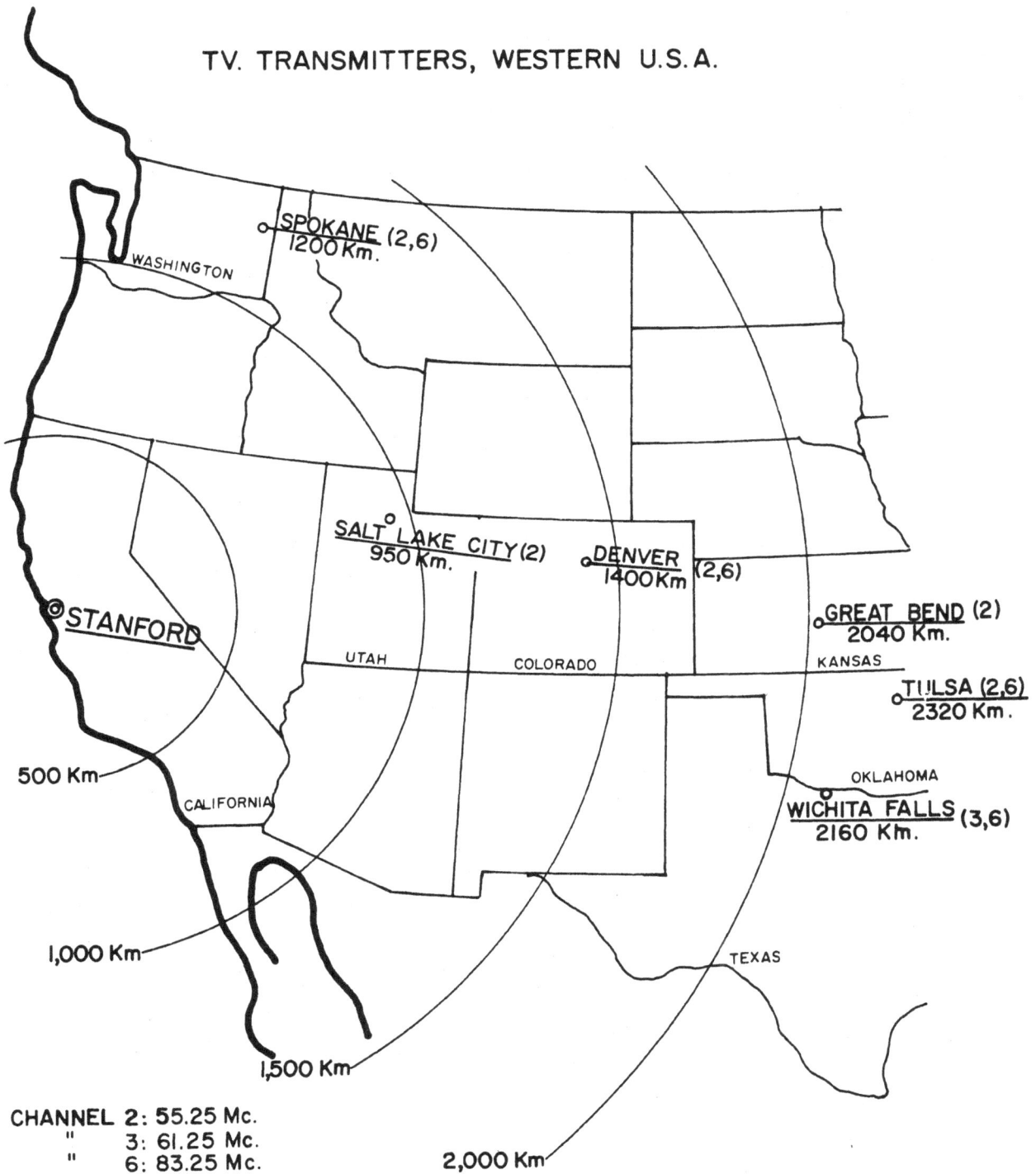

TV. TRANSMITTERS, WESTERN U.S.A.

SPOKANE (2,6)
1200 Km.

WASHINGTON

SALT LAKE CITY (2)
950 Km.

DENVER (2,6)
1400 Km.

GREAT BEND (2)
2040 Km.

◎STANFORD

UTAH   COLORADO   KANSAS

TULSA (2,6)
2320 Km.

500 Km

CALIFORNIA

OKLAHOMA

WICHITA FALLS (3,6)
2160 Km.

1,000 Km

1,500 Km

TEXAS

CHANNEL 2: 55.25 Mc.
   "   3: 61.25 Mc.
   "   6: 83.25 Mc.

2,000 Km

FIG. 32.--Locations of the television stations used in the tests
of echo-duration dependence on path length and wavelength.

approximately 12 feet above ground. The video carrier (55.25 ± 0.01 Mc) amplitude was recorded on two direct-inking pen-motors fed by the output of two receivers having 6-kc band-width. The gains of the two penmotor channels were adjusted differently so as to increase the dynamic range of the recordings.

The results of the duration-dependence-on-path-length tests are summarized in Table IV. Note that no meteor-propagated signals were recorded from Tulsa, a distance of 2320 km from Stanford. On the other hand, more than a hundred echoes per hour (including non-expoential echoes) were recorded from Great Bend, a distance of 2040 km. The transmitting facilities at Tulsa and Great Bend are similar. In a previous three-hour test with Tulsa (February 28, 1956), negative results were also obtained. In similar tests with two stations (channels 3 and 6) in Wichita Falls, Texas, on August 10, 1956, no signals were received though the frequencies were monitored for about three hours in the early morning. The distance to Wichita Falls from Stanford is 2160 km. Thus it appears that the maximum distance over which meteoric propagation is possible is rather sharply defined. For the receiving arrangement at Stanford, meteor propagation at a burst rate of more than 100 per hour was obtained with a transmitter at 2040-km distance, while no signals could be obtained from similar transmitter sites at ranges of 2160 and 2320 km.

Only the short-duration echoes (presumably from low-density trails) were scaled in the duration-dependence tests, since the theoretical effects of wavelength and path length on duration are better understood for low-density trails than for high-density trails. In an effort to make the measurements independent of the threshold (noise) level, the decay time-constants (intrinsic durations) of the echoes were measured instead of their total observed duration. Most of these echoes appeared to decay approximately exponentially, though this factor was not investigated thoroughly. Assuming that the decay was exponential (low-density, long-wavelength theory), the decay time constant was determined from measurements

TABLE IV. Duration dependence on distance.

	Station KUTV Salt Lake City Utah	Station KTVR Denver, Colo.	Station KCKT Great Bend Kansas	Station KVOO Tulsa Oklahoma
Video Carrier frequency	55.24 Mc	55.25 Mc	55.25 Mc	55.26 Mc
"Effective" Radiated Video Power	46 kw	100 kw	100 kw	100 kw
Date and time of test	3/31/56 0300-0400 PST	3/30/56 0200-0300 PST	3/15/56 0145-0315 PST	3/24/56 0100-0400 PST
Distance from Stanford	930 km	1500 km	2040 km	2320 km
Bearing from Stanford	65°	75°	78°	84°
Number of exponential echoes recorded	320	200	115	none
Average intrinsic echo duration $\lambda^{2}sec^{2}\emptyset/32\pi^{2}D$	0.123 sec	0.258 sec	0.221 sec	---

at two points on the echo separated as far as possible in intensity. The intrinsic durations determined experimentally should be compared with $(\lambda^2 \sec^2\phi)/32\pi^2 D$, assuming that the low-density, long-wavelength theory is applicable (see Table II and associated discussion).

In Fig. 33, the experimental intrinsic echo durations noted in Table IV are plotted as a function of path length. The point at zero distance ($\tau_o = 0.0233$ seconds) was computed assuming $D = 4m^2/s$, an average value determined from radar measurements. This value of diffusion coefficient is believed to apply at approximately 93 km height.[20] The point at zero distance is connected to the experimentally-determined points by a line which is dotted to emphasize that it is not firmly established by the few experimental points.

Note from Table IV and Fig. 33 that the experimental echo durations first increase with path length, and then decrease as the length of the path is increased beyond approximately 1500 km. Assuming a constant height (constant D), the simple first-order theory predicts that echo duration in these tests should have varied as $\sec^2\phi$, where 2ϕ is the angle between rays from the meteor to the transmitter and receiver. The position of each meteor was not determined in these tests, but an upper and lower limit for $\sec^2\phi$ can be determined from the geometry of the paths. The maximum $\sec^2\phi$ occurs for the midpoint of the path, while the minimum occurs when the take-off angle for the ray at one end of the path is zero, and the bearing of this ray is toward the other end of the path. In the figure, the solid lines marked (max $\tau_o \sec^2\phi$) and (min $\tau_o \sec^2\phi$) were computed assuming that the back-scatter intrinsic echo duration $\tau_o = 0.0233$ seconds, and assuming a constant height of approximately 100 km in the computation of the maximum and minimum values of $\sec^2\phi$. Comparison of the experimental points with these theoretical curves indicates that for path lengths greater than perhaps 1200 km, the echo durations are appreciably less than anticipated from the first-order theory.

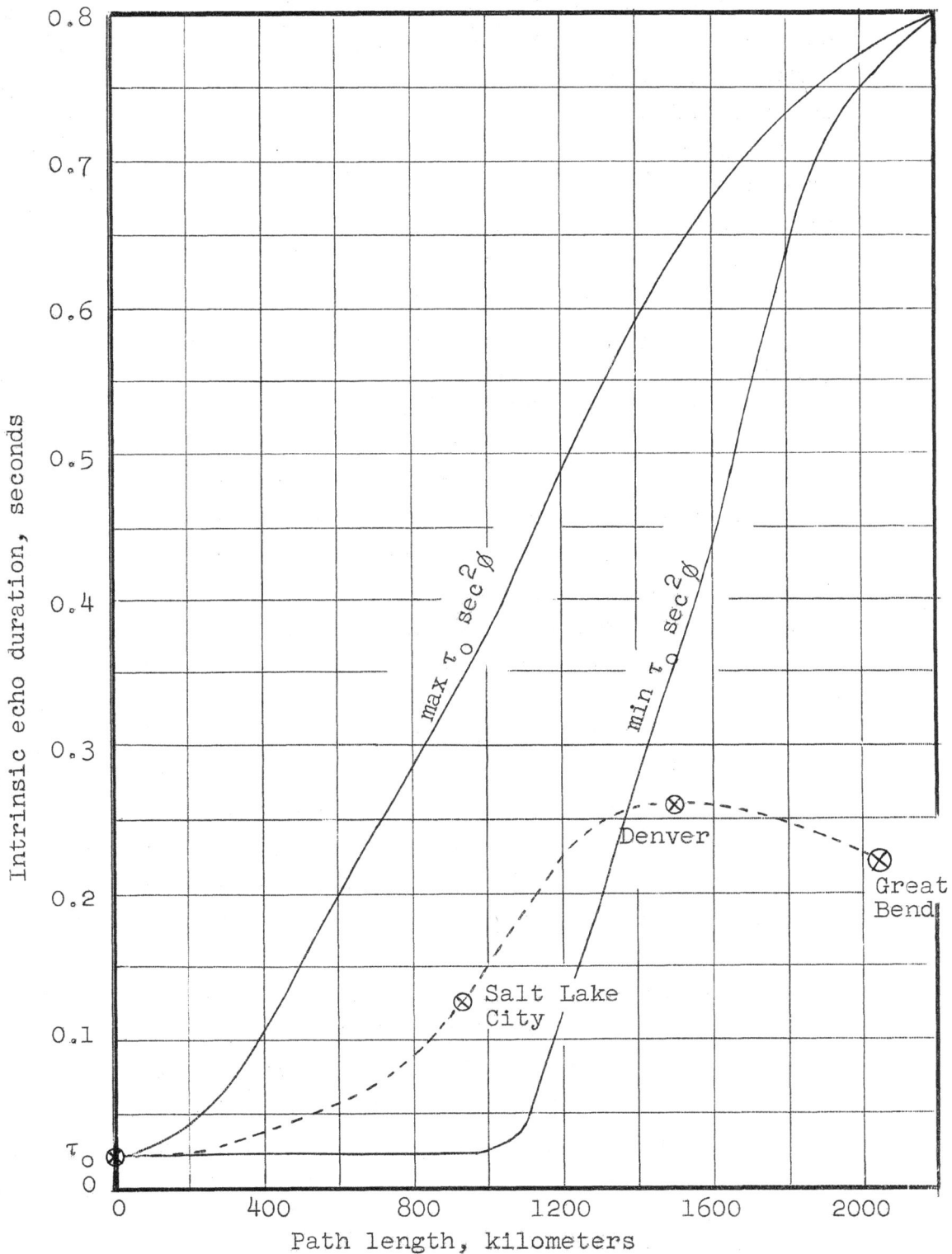

FIG. 33.--Intrinsic echo duration plotted as a function of path length. The maximum and minimum theoretical values (solid line) are based on the assumption of a constant mean height. The measured values are on the dotted lines.

The noted trend from theory of the experimental durations had been anticipated for the following reasons.* (1) The effects of turbulence in changing the shape of the trail, and the effects of electronic recombination and attachment in changing the line density, are both expected to make the exponent of λ and sec $\emptyset$ in the duration dependence less than two. The amount by which this exponent falls below two is expected to increase with path length, and with increased echo duration. (2) As the path length for meteor propagation is extended toward its maximum value, it is expected on geometrical arguments--based on the antenna patterns--that the mean height of the echo-producing meteors will increase. Since the diffusion coefficient D increases rapidly with height (and echo duration is inversely proportional to D), this effect might become more important than the increasing values of $\sec^2 \emptyset$ when path length is increased beyond some fairly long value. For very long paths, therefore, echo duration would decrease with increasing path length.

It appears that the above explanations can account qualitatively for the observed durations. In connection with the factors discussed in (1) above, it is important to note that the exponents of λ and sec $\emptyset$ are expected to change in the same manner. In the tests described thus far, λ was held constant and $\emptyset$ was changed by changing path length. Another measure of some of the same factors which control echo duration can be made by holding path length constant and varying λ.

On the morning of August 14, 1956, stations KTVR (55.25 Mc) and KRMA (83.24 Mc) in Denver, Colorado, were simultaneously monitored at Stanford. In about two and a half hours of operation, exponential echoes were simultaneously recorded on both frequencies for 63 meteors. Upon scaling the decay time constants and comparing them with the radio wavelengths for the

---

*See Quarterly Status Report No. 8, February 20, 1956.

two stations, it was found that if the echo durations are proportional to λ^n, then n = 0.851. When the records were scaled for total echo durations at 3 db below peak echo intensity, it was found that n = 1.24.

On the morning of August 23, 1956, a similar test was held with stations KREM (55.24 Mc) and KHQ (83.24 Mc) in Spokane, Washington. For 75 simultaneously-recorded short-duration echoes, n = 1.07 based on the decay time constant, and n = 1.42 based on total echo durations at 3 db below peak echo intensity. It is difficult to determine the significance of the measured difference in the exponent for the approximately N-S and E-W paths, due to the paucity of data. As mentioned previously, an attempt was made to monitor channels 3 and 6 in Wichita Falls, Texas (2.60 km distance), but no echoes were received.

The results of the duration-dependence-on-wavelength tests show the same type of deviation from the theory as do the distance tests. Since the $\lambda^2 \sec^2 \phi$ factor is a fairly good approximation for radar (back-scatter) and short-path measurements, these deviations must be associated with changes which occur only for relatively long paths (say 500 km and greater). In addition to the possible explanations given previously, it may be that, at the shortest wavelengths (3.6 meters) used in the wavelength dependence tests, the echoes displayed some of the characteristics associated with the low-density short-wavelength theory. While durations are proportional to $\lambda^2 D^{-1} \sec^2 \phi$ for long wavelengths, these factors affect duration quite differently at the short wavelengths. For $\lambda < \lambda_T$, echo duration is proportional to $\lambda^{-1} D \sec^{-1} \phi$ for meteors perpendicular to the propagation path, while they are proportional to $\lambda^{-1} D \sec \phi$ for meteor trails which lie along the path. For oblique propagation, λ_T is proportional to $D^{2/3} \sec^{-1} \phi$ for across-the-path trails, and to $D^{2/3} \sec^{-1/3} \phi$ for along-the-path trails. If mean height increases fast enough for the echo duration to decrease with path length, it will also cause the transitional wavelength between long and short-wavelength theory to occur at longer wavelengths as path length is increased. Thus the measured exponents of λ may result from a mixture of 2 from long-wavelength theory and -1 from short-wavelength theory. Similarly, the departure of the measurements from the $\sec^2\emptyset$ theory could result from a mixture of echoes where the $\sec^2\emptyset$ law applies, and others where the duration is proportional to $\sec\emptyset$ or $\sec^{-1}\emptyset$.

While the various explanations offered above can be used to explain qualitatively the measurements of echo duration dependence on path length and wavelength, the total data are too limited to determine the precise role played by each of these possibilities. However, the last mentioned effect (that some of the echoes may correspond to the short-wavelength theory) is not believed to be as important in the explanation of the results as the expected change in average meteoric height for the longer paths, and the effects of turbulence, recombination, and attachment. This conclusion is based upon the observed intensity-time characteristics of the echoes, although no qualitative measure based on these characteristics has yet been applied to test this conclusion. Because of the importance of echo duration in meteor-burst propagation, it is important that more extensive experiments of the types discussed here be carried out.

A preliminary account of the echo duration dependence on wavelength and path length was presented at the symposium on "Current Problems in Radio Wave Propagation" in Paris, Sept. 17-22, 1956. The presentation was entitled "The effect of propagation path-length on the duration of meteor echoes," and was authored by V. R. Eshleman and A. M. Peterson. It was presented by V. R. Eshleman.

Other experimental results on the dependence of echo duration on path length and wavelength have been reported. McKinley and McNamara[21] recorded meteor bursts simultaneously at back-scatter and over a 338-km path in Canada at a wavelength of 9 meters. They found the mean value of n in the

$\sec^n \phi$ dependence to be approximately 1.73 for short-duration echoes, and 1.13 for long-duration echoes. Forsyth and Vogan[22] found from measurements on 6 and 8 meters over 1000-km N-S and E-W paths in Canada that the n in an assumed λ^n dependence may be more or less than two. For the short-duration echoes on the E-W path, the mean value of n was found to be 2.3, while for the N-S path it was 1.6. (It was suggested that the difference between the two paths might be due to the differential effect of the radiant distribution on the two paths, and/or the effect of the geomagnetic field.) It is important in comparing these data with the results of the TV recording tests to note the differences in path length, radio wavelength, and latitude of the paths. A satisfactory explanation of all of the experimental results is not yet available.

B. INTENSITY AND DURATION DISTRIBUTIONS OF METEOR ECHOES
   PROPAGATED OVER A LONG PATH

The number-intensity and number-duration distributions of meteor echoes have previously been studied at various frequencies for back-scatter and for several forward-scatter paths as long as 1000 km. In this section, the intensity and duration distributions of the meteor signals propagated over a 2040 km path are presented. This is the path from Great Bend, Kansas to Stanford, California which is described in the preceding section on echo durations.

The number-intensity distribution is presented in Fig. 34. The logarithm of the number N of echoes of peak intensity P_R or greater is plotted as a function of log P_R, where P_R is in milliwatts at the receiver input. A total of 181 echoes were scaled for these results. Straight lines are shown in the figure for N proportional to $P_R^{-0.5}$, $P_R^{-1.0}$, and $P_R^{-2.0}$. It is apparent that the measured results are closely described by N proportional to $P_R^{-1.0}$.

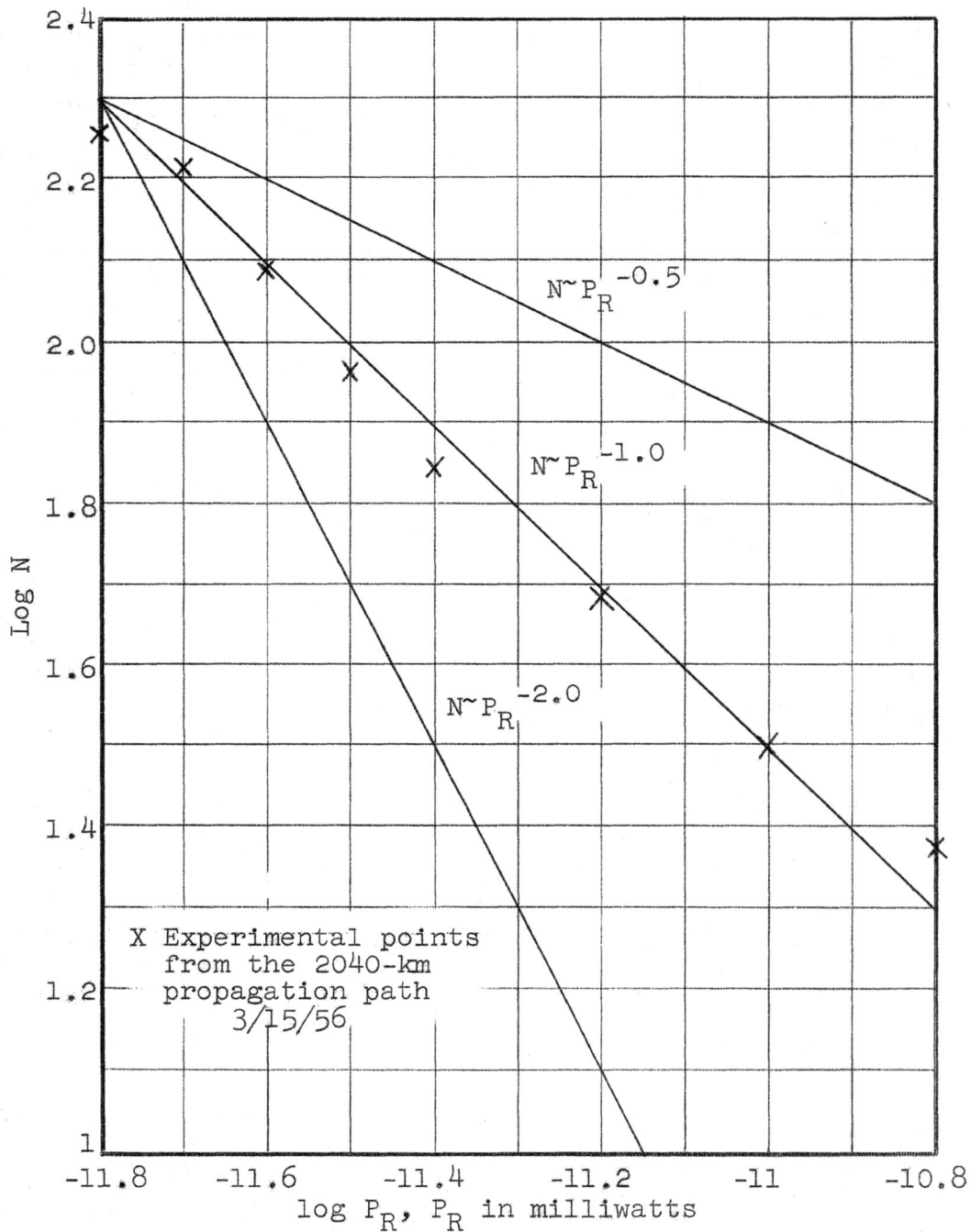

FIG. 34.--The log of the number of echoes of intensity greater than P_R plotted as a function of log P_R; theoretical curves and experimental points.

In the approximate low-density theories for both long and short wavelengths, P_R is proportional to q^2. From the simple theory for high-density trails at long wavelengths, $P_R \sim q^{1/2}$. It has been determined from radar measurements that the number of trails of line density q or greater is proportional to $q^{-1.0}$. Thus, from the simple theories, N should be proportional to $P_R^{-0.5}$ for low-density trails, and to $P_R^{-2.0}$ for high-density trails for long wavelengths. (It is believed that the various parameters for this long-path are such that $\lambda > \lambda_T$.) Radar measurements on the intensity distribution of echoes show a transition between the low- and high-density regions, with the asymptotic values corresponding to those predicted above.

The measurements on the long path show a dependence of N on P_R which is intermediate between the extremes observed for back-scatter. Over the 10 db variation in intensity of the measurements, however, there is no apparent curvature which might show that the slope of the curve is changing to -0.5 at one end or -2.0 at the other. In fact, other measurements of echo-intensity distributions for forward scatter (500 to 1000 km paths) over intensity ranges of as much as 30 db, indicate that one straight line can be made to fit the experimental results very well.[23,24,25] The slopes reported by various investigators for these shorter paths vary between -0.7 and -1.1. The exponent of P_R appears to change diurnally, and it is also probably a function of season and path geometry. On the basis of the one curve on the long path, it does not seem likely that there are large or important differences in the number-intensity distributions for long and intermediate path lengths.

A qualitative explanation is offered below for the observed linear relationship between number and intensity for the oblique paths, as opposed to the observed change in slope for back-scatter. In the low-density regions, the amplitude distribution should be the same with forward scatter as with back scatter.

For high-density trails and forward scatter, the number is expected to decrease less rapidly with increasing P_R than for back scatter because of the formation of the geometrical-optics shadow-zone lobe near the forward direction (see Figs. 3 and 4). In fact, in the exact forward direction, the number-intensity relationship for high-density trails should be the same as for low-density trails. This follows from the argument that when the trail expands to the point where there are no critical density regions, all of the electrons in the principal Fresnel zone scatter in phase in the exact forward direction. In addition to these fundamental differences in the high-density region between back and forward scattering, it would be expected that the transition between high- and low-density regions would be more extensive for forward scatter than for back scatter, with the measurement techniques that have been used. For instance, the range to the meteor is easily measured and used in correcting measured amplitudes to a standard range for back scatter, while in forward scatter, the ranges from transmitter to trail and from trail to receiver have not been measured in the reported intensity-distribution tests. Also, there are more variables controlling echo amplitude for forward scatter than for back-scatter. If the above factors are responsible for the measured slopes being intermediate to -0.5 and -2.0, with little change in slope over a fairly-wide range of intensities, it would be expected that a change in slope would be found if measurements were made over a much wider range of intensities. In particular, it is believed that the slope would approach -0.5 for a high-sensitivity system capable of detecting very low-density trails.

In order to analyze the number-duration distribution of meteor echoes, it is not convenient to use the intrinsic duration (τ) since this factor is not defined for the long-enduring echoes from high-density trails. Instead, the total duration (d) above the noise level will be used. When the logarithm of the number of echoes of duration greater than d is plotted as a function of log d, the curve shows an inflection or change in slope. Such a behavior is fairly unusual in physical phenomena, but in this case it is just what is needed to correspond to the theoretical predictions. Such a plot for 170 echoes obtained over the 2040-km path is presented in Fig. 35. The crosses are the experimental points, and the solid lines are obtained from the theory.

In section IIB it was shown that $d = \tau \ln(q/q_o)^2$ for long-wavelength and low-densities. For high densities, we see from Table II that echo duration is proportional to q. Since the number of trails of line density greater than q is proportional to $q^{-1.0}$, the number N of trails of duration greater than d is proportional to $\exp(-d/2\tau)$ for low-densities and to d^{-1} for high-densities. These two expressions are the theoretical solid lines in Fig. 35. They have been translated in both the ordinate and abcissa directions to fit as closely as possible with the experimental results. The position for the fit of the $\exp(-d/2\tau)$ curve with the experimental points indicates that $\tau = 0.132$ seconds. This value of τ is considerably less than the value (0.221 sec) scaled directly from the short-duration echoes of the same experiment. Thus, while it is possible to match up the experimental points with the theoretical curves, it does not appear that this close agreement necessarily means that the experimental results are in complete accord with the theory. More work along these lines is needed. A closer investigation of the decay characteristics of the echoes may help solve some of the remaining questions on echo-duration characteristics.

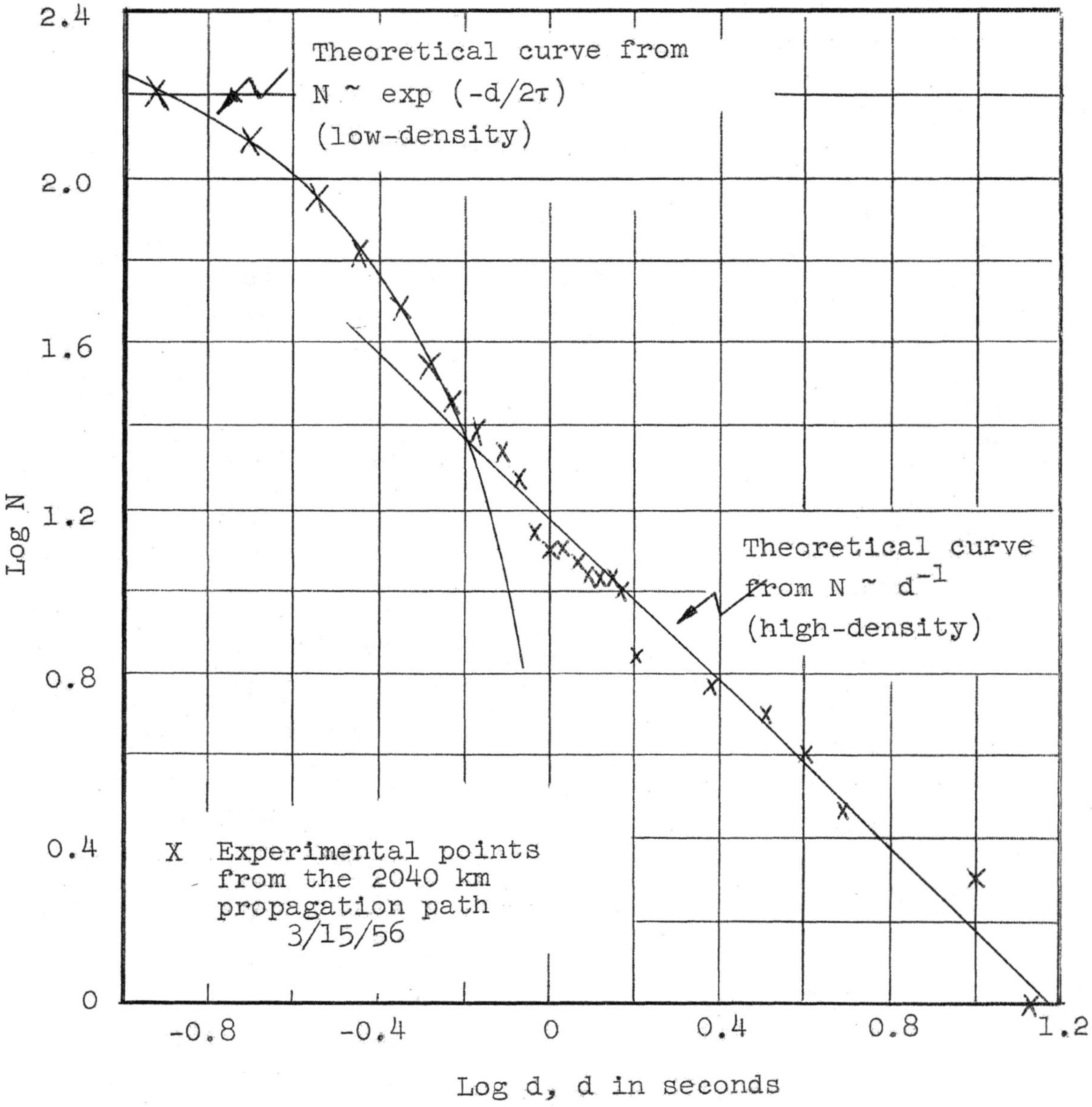

Theoretical curve from
N ~ exp (-d/2τ)
(low-density)
Theoretical curve
from N ~ d-1
(high-density)
X Experimental points
from the 2040 km
propagation path
3/15/56
Log N
Log d, d in seconds
2.4
2.0
1.6
1.2
0.8
0.4
0
-0.8
-0.4
0
0.4
0.8
1.2

FIG. 35.--The log of the number of echoes of duration greater than d plotted as a function of log d; theoretical curves and experimental points.

VII. MEASUREMENTS OF VERY SMALL METEORS

Previous radar investigations of the ionized trails produced
by meteoric particles extend down to the small meteors corresponding to an estimated 10th visual magnitude (where the electron
line-density is 10^{12} electrons per meter). The experimental
study of yet smaller ionizing meteors (down to micrometeorite
sizes) had not been possible due to the lack of sufficiently
sensitive research tools. It appears that there is a gap in
the spectrum of meteoric sizes about which little is known.
The width of this gap is estimated at 8 visual magnitudes, or a
range of about 10^3 in mass and electron line-density. The
object of the phase of the meteor program described here is to
extend the radio investigation of meteors as far as possible into this "missing link" of the spectrum.

The construction of a more sensitive meteor radar, capable
of detecting ionization produced by meteors a hundred-fold less
massive than heretofore possible, was undertaken during the
latter half of the contract year 1956. With the system partially
completed and in operation, initial results have become available for interpretation.

The main objectives of this study fall into two categories:
(1) to learn about certain astronomical aspects of these small
particles; for example, their velocities, numbers, and origin
(interstellar or interplanatary); (2) to determine their contribution to characteristics of the earth's upper and outer
atmosphere, such as E-region ionization, F-corona, zodiacal
light, and the gegenshein. In particular, the characteristics
of the smallest meteors which create ionization, and their role
in meteor burst and ionospheric scatter propagation, are important questions to be answered.

The experimental installation features a high-gain, double-
line broadside array of 96 four-element Yagi antennas designed
to operate on 23.1 Mc. This array is pictured in Fig. 36. The
23.1 Mc frequency was chosen as a compromise between smaller

FIG. 36.--Photograph of 23.1 Mc double-line high gain array looking west. 48 4-element Yagis spaced one wavelength apart in each row are each mounted on wood telegraph poles at heights above ground varying from 25 to 52 feet. Meteor radar is housed in building at far end.

echoing power at higher frequencies, and ionospheric layer and backscatter interference at lower frequencies. Design studies, involving cost analysis and mocrowave model measurements, were undertaken by the Stanford Research Institute, Menlo Park, California, under separate contract with A.F.C.R.C. The individual Yagi antennas (see Fig. 37) have a measured 10 db gain over an isotropic radiator, and a minimum 25 db front to back ratio. Each of the two rows contain 48 Yagi antennas. The length of each row is 2010 feet (48 wavelengths), and the two parallel rows are separated by 320 feet (7.5 wavelengths). Assuming negligible mutual impedance between adjacent antennas (separated by one wavelength), the theoretical gain of the antenna is $10 + 3(\log_2 96) = 29.8$ db, relative to an isotropic radiator. The actual gain of the antenna has not yet been measured.

The area selected for the installation of the array is on the Stanford campus. The site choice involved a compromise among several factors; low noise level, minimum initial and maintenance cost of the array, easy access to the site, and permanence of the installation. Although it was desired to orient the array on a horizontal east-west line and thereby concentrate radiation in the north-south meridian plane, the slope of the terrain at this best available location requires that the plane normal to the line of the array be shifted west 3.5^O at the zenith and east 2.5^O in azimuth at the horizon.

The theoretical radiation pattern produced by each line of the (48) Yagi antennas in the array is a fan-shaped beam directed normal to the line. In the elevation plane, the pattern is that of a single 4-element Yagi. The center of the pattern may be directed to different north or south elevation angles by manually tilting each antenna of the array. The theoretical transverse pattern pictured in Fig. 38 indicates an expected half-power beamwidth of 1.5 degrees for each row of antennas. Experimental verification of this value has been obtained utilizing recorded emissions from the radio noise

FIG. 37.--Single 4-element rotatable Yagi antenna which can be used for auxiliary studies in conjunction with high gain array. Gain is 10 db with minimum front-back ratio of 25 db.

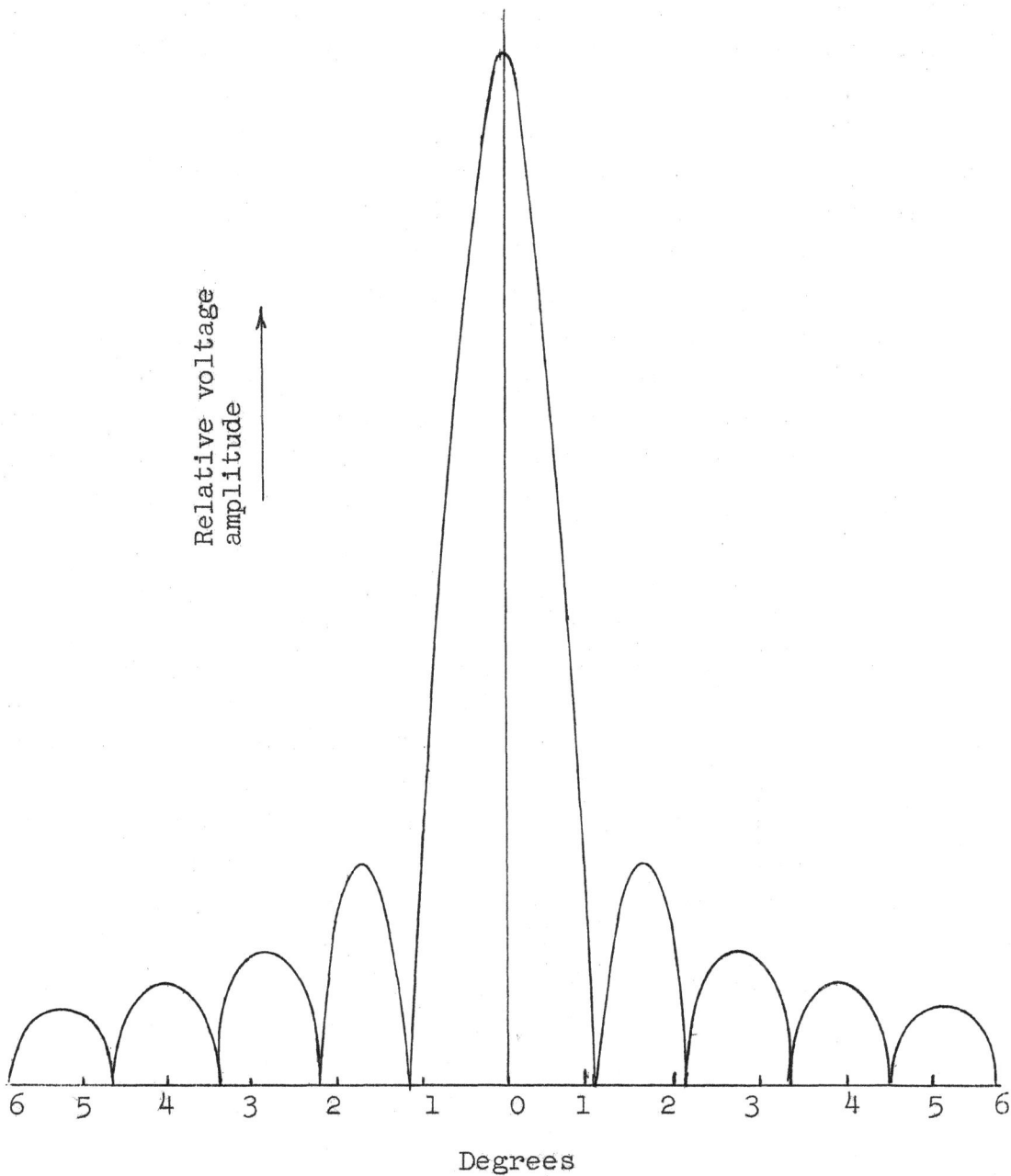

FIG. 38.--Calculated transverse pattern for a single line
of 48 four-element Yagis spaced one wavelength apart.
Half power beamwidth is 1.2 degrees.

source Cassiopeia (see Fig. 39). When both rows operate simultaneously, the transverse pattern is narrowed only slightly, and the elevation pattern assumes a 7 lobe structure. This lobe structure will be useful for angle of arrival measurements.

The feed arrangement features balanced, open-line directional couplers at each antenna (see Figs. 40 and 41). These are designed to couple to each antenna equal amounts of power from the 450 ohm balanced feedline which runs along each row of poles. In-phase power from the 475 ohm output of each directional coupler is matched to the balanced 200 + j0 ohms antenna impedance through a 300 ohm quarter-wave matching section located at the coupler. A length of balanced 200 ohm transmission line (Federal K-200 1 kw) is used between this point and the antenna terminals. This feed system has been designed to handle at least 200 kw peak pulse power with 1 db loss in each row of antennas.

The meteor transmitter--a hybrid unit on loan from another contract of the Stanford Radio Propagation Laboratory--is capable of 90 kw peak pulse power at 3% maximum duty cycle. A bandwidth of 10 kc is used in the receiver, this bandwidth being a compromise between maximum sensitivity and the resolution required to distinguish individual meteoric echoes without excessive overlapping. The complete radar system should detect trail line-densities of 5×10^9 electrons per meter (approximately 15th magnitude meteors) at a 200 km range.

With one half of the array completed and in operation, trail densities to about 10^{10} electrons per meter have been detected and recorded as individual events. In Fig. 42 is shown a 40-second sample of strip film range-time recording of the meteoric activity during the early morning hours. The influx rate of favorably-oriented trails during these hours is as high as 5000 per hour. Most of the echoes are of short (less than one second) duration.

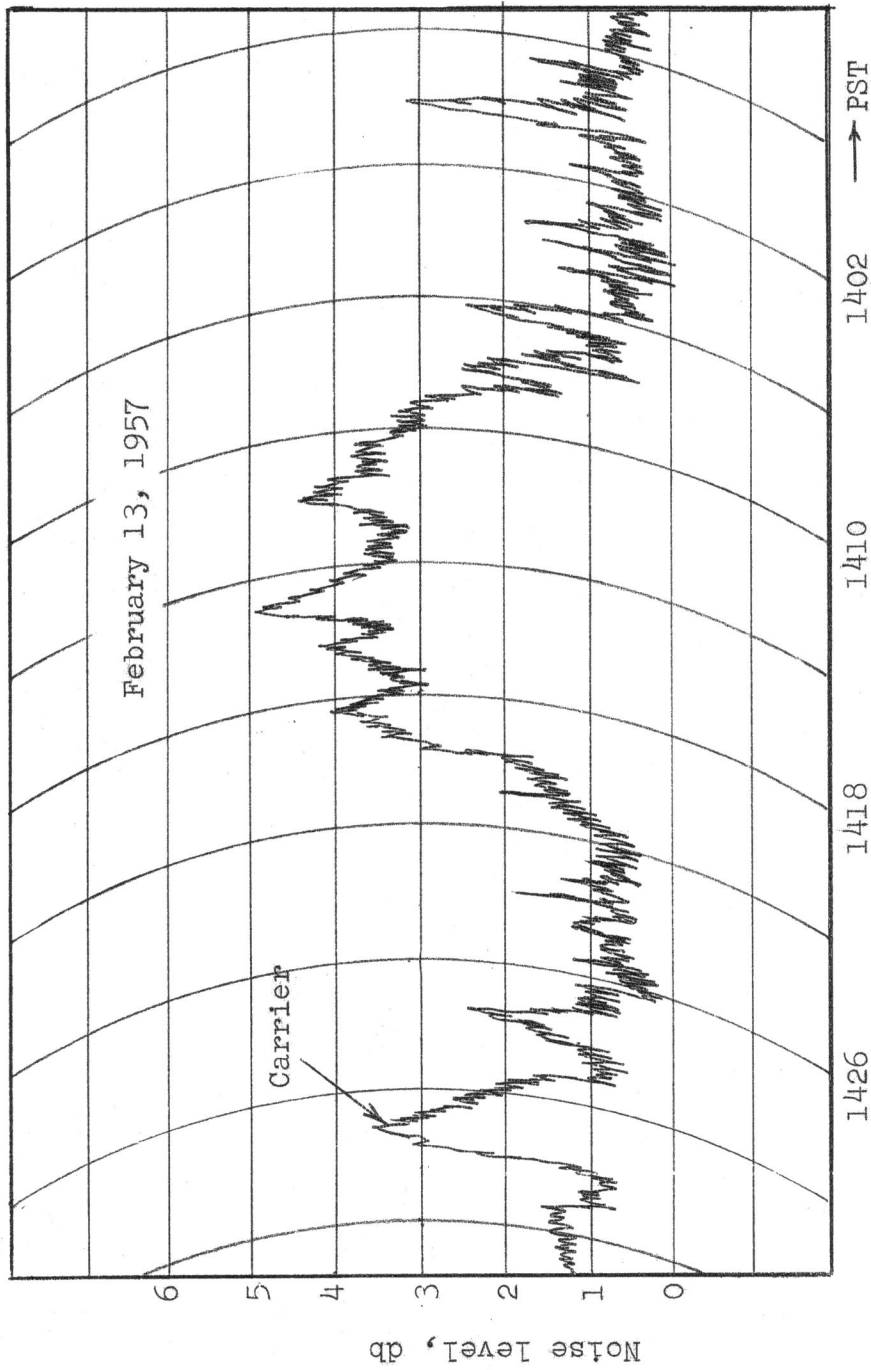

FIG. 39.--Penmotor recording of the radio noise source Cassiopeia crossing the beam generated by one row of 48 Yagi antennas operating on a frequency of 23.1 Mc. Each four-minute division represents one-half degree for this source which transits at an elevation angle of 67.5 degrees.

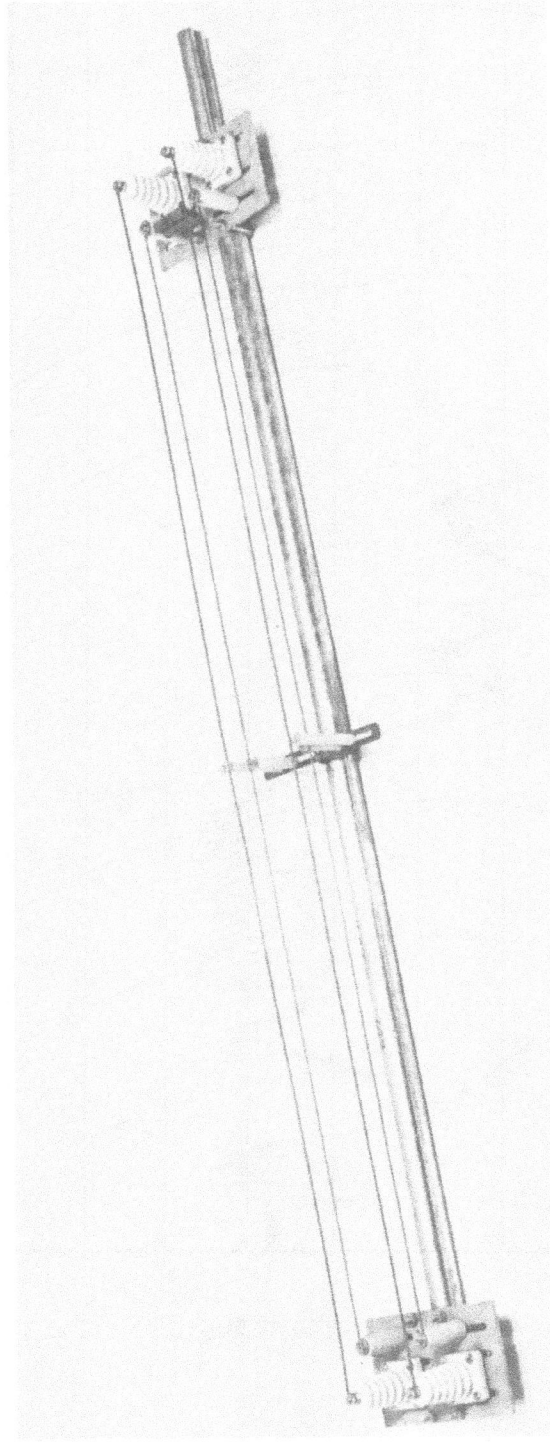

FIG. 40.--Photograph of directional coupler prototype used at each antenna. Upper pair (on ribbed insulators) are a part of the 450 ohm main feed line. Antenna is fed at left end of 475 ohm secondary (lower pair of wires); a 475 ohm dummy load terminates other end. Useful power is coupled to and from the left. Lengths vary from 4 to 10 feet depending on where it is inserted in the array.

FIG. 41.--Section of main feed line along one row of poles (looking east) showing couplers installed.

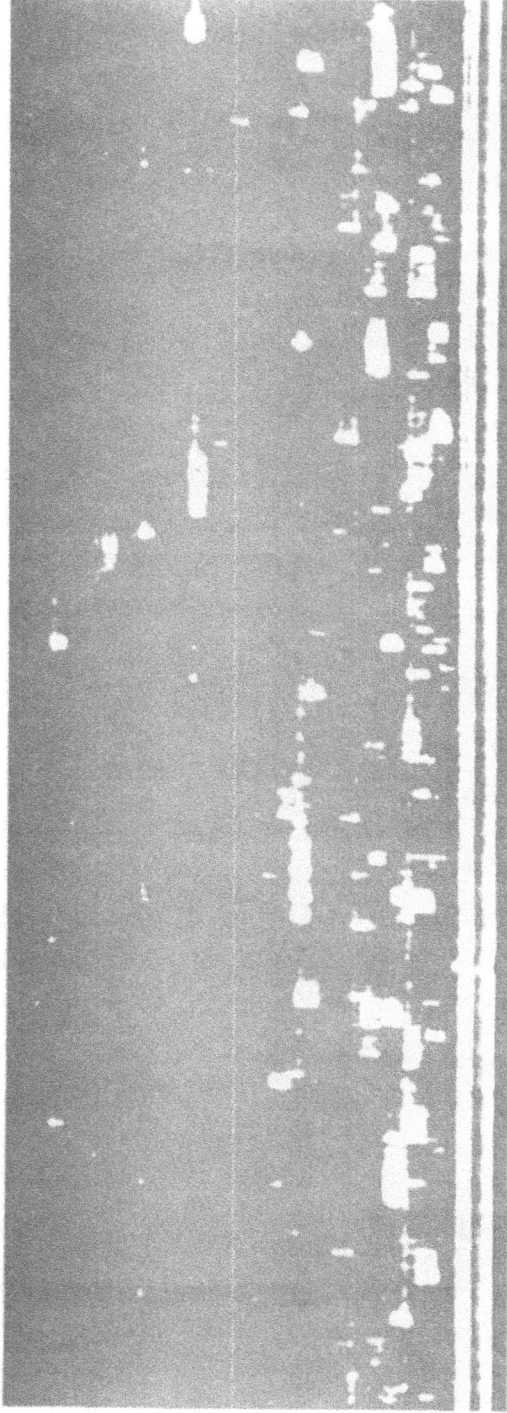

FIG. 42.--Range-time display of meteor activity recording at Stanford using the 48-Yagi 23.1-Mc array beamed north; 100 km range marks; peak power output 35 kw; January 24, 1956; 0130 PST. (Blown-up forty-second sample).

Initial measurements have been made in an effort to establish the trend of the rate-amplitude law for the echoes from very small meteors. Data were taken during the early morning hours when the maximum influx of meteoric particles occurs. These new data indicate that the number of echoes from trails of line-density greater than q is inversely proportional to q when q is greater than about 2×10^{10} electrons per meter. However, below this value of line density, it appears that the number of small meteors is even larger than would be predicted by an extrapolation of this relationship. An example of one set of rate-amplitude measurements is shown in Fig. 43.

The antenna array, when completed this comming summer, will render the system sensitive to trail densities of about 5×10^{9} electrons per meter. In addition to the meteor investigations, other studies related to weak ground-backscatter echoes and ionization originating at E and F region heights, will be conducted. The disposition of the array is favorable also for such radio astronomy studies as emissions from the sun, planets, and stellar sources at a wavelength of 13 meters.

It appears that a further increase in the radar sensitivity will be required to complete the investigation of meteors to the smallest ionizing size. The transition region from ionizing particles to the smaller nonionizing micrometeorites is apparently still beyond the sensitivity of the present apparatus. The remaining uncharted meteors may have an important role in meteor communication and in various radio astronomical events. The logical next step in increasing sensitivity is to increase the power of the radio transmitter.

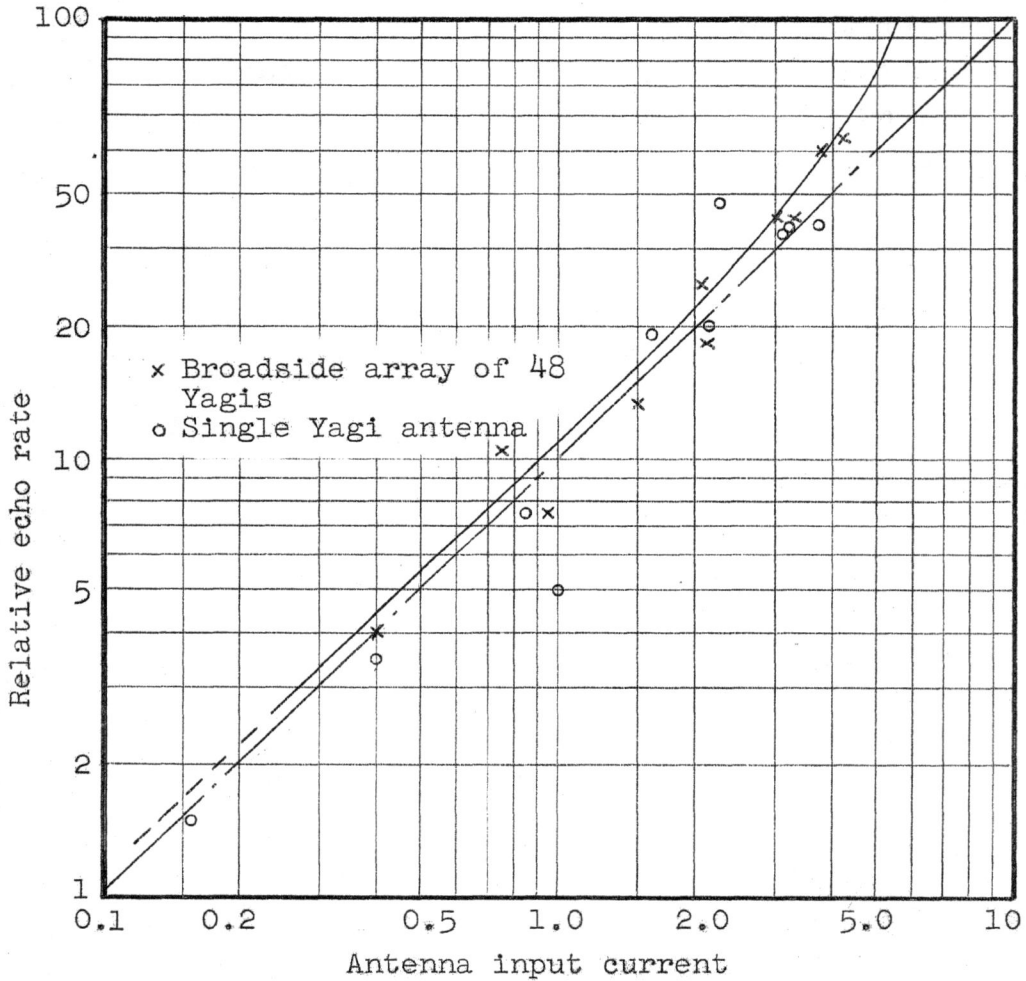

FIG. 43.--Plot of (normalized) meteor count versus antenna input current from experimental data taken at Stanford, California on 23.1 Mc using one line (48 antennas) of the high gain array. February 28, 1957; 0400 - 0450 PST; peak power output, 90 kw.

VIII. CONCLUSIONS AND RECOMMENDATIONS

While much is now known about meteoric phenomena, much remains to be done. More detailed studies of the type performed under this contract and elsewhere, are needed for a complete understanding of meteors. We believe that future theoretical and experimental research on meteors and meteoric ionization should be concentrated in the following areas: (1) studies of very small meteors; (2) studies of the effects of wavelength and path length on echo characteristics; (3) investigations of the fine-scale variations in the meteor radiant distribution; and (4) investigation of short-wavelength echo characteristics.

The expected applications of such researches are: (1) to help determine the role played by meteors in ionospheric-scatter propagation. Studies on small meteors are also needed to determine the relationship between meteors and scattered light from the sun (F-corona, zodiacal light, and gegenshein). An assessment of the total effect of meteors requires knowledge over the entire spectrum of meteor sizes. (2) Detailed studies of meteor echoes propagated over various paths using various radio wavelengths, have important applications in the determination of the characteristics of the upper atmosphere, and the characteristics of meteor propagation. (3) Information on the origin and space distribution of meteoric material could be obtained from a more careful study of meteor radiants. The fine-scale directional characteristics of meteor propagation, and the feasibility of a prediction service for meteor propagation, could be determined from full information on meteor radiants. (4) Short-wavelength studies are needed to determine the highest feasible frequencies for meteor-burst propagation, and for ionospheric-scatter propagation. Research on short-wavelength echoes should also yield new information on the size and effect of the initial trail radius, the rate of expansion of the ionization, and the prevalence of fragmenting meteors.

BIBLIOGRAPHY

1. D. K. Bailey, R. Bateman, L. V. Berkner, H. G. Booker, G. F. Montgomery, E. M. Purcell, W. W. Salisbury, and J. B. Wiesner, "A new kind of radio propagation at very high frequencies observable over long distances," Phys. Rev., 86, 141 (1952).

2. O. G. Villard, Jr., A. M. Peterson, L. A. Manning, and V. R. Eshleman, "Extended-range radio transmission by oblique reflection from meteoric ionization," Jour. Geophys. Res., 58, 83 (1953).

3. V. R. Eshleman and L. A. Manning, "Radio communication by scattering from meteoric ionization," Proc. IRE, 42, 530 (1954).

4. D. W. R. McKinley, "Dependence of integrated duration of meteor echoes on wavelength and sensitivity," Can. Jour. Phys., 32, 450 (1954).

5. P. A. Forsyth et. al., "Project Janet," various Radio Physics Laboratory Reports, Defense Research Telecommunications Establishment, Ottawa, Canada.

6. V. R. Eshleman, "The mechanism of radio reflections from meteoric ionization," Tech. Rept. 49, Electronics Research Lab., Stanford Univ., Calif. (1952).

7. N. Herlofson, "Plasma resonance in ionospheric irregularities," Arkiv for Fysik, 3, 247 (1951).

8. T. R. Kaiser and R. L. Closs, "Theory of radio reflections from meteor trails: I," Phil. Mag., 43, 1 (1952).

9. R. E. B. Makinson and D. M. Slade, "Dipole resonant modes of an ionized gas column," Aust. Jour. Phys., 7, 268 (1954).

10. V. R. Eshleman, "The effect of radar wavelength on meteor echo rate," Trans. IRE, AP-1, 37 (1953).

11. T. R. Kaiser, "Radio echo studies of meteor ionization," Research Reviews, Phil. Mag. Supplement, 2, 495 (1953).

12. M. Loewenthal, "On meteor echoes from under dense trails at very high frequencies," presented at the joint URSI-IRE meeting in Washington, D. C., May 1, 1956. To be published as a Lincoln Laboratory Technical Report.

13. G. S. Hawkins, "Radar echoes from meteor trails under conditions of severe diffusion," Proc. IRE, 44, 1192 (1956).

14. W. A. Flood, "Meteor echoes at ultra high frequencies," private communication.

BIBLIOGRAPHY (Cont'd)

15. L. A. Manning, O. G. Villard, Jr., and A. M. Peterson, "The length of ionized meteor trails," Trans. Am. Geophys. Un., 34, 16 (1953).

16. E. Eastwood and K. A. Mercer, "A study of transient radar echoes from the ionosphere," Proc. Phys. Soc., 61, 122 (1948).

17. G. S. Hawkins, "A radio echo survey of sporadic meteor radiants," M.N.R.A.S., 116, 92 (1956).

18. O. G. Villard, Jr., V. R. Eshleman, L. A. Manning, and A. M. Peterson, "The role of meteors in extended-range uhf propagation," Proc. IRE, 43, 1473 (1955).

19. V. R. Eshleman, "Meteors and radio propagation; Part A," TR No. 44, Applied Electronics Laboratory, Stanford Univ., Stanford, California, February, 1955.

19a. C. O. Hines, "Diurnal variations in the numbers of shower meteors detected by the forward scattering of radio waves," Can. Jour. Phys., 33, 493 (1955).

19b. C. O. Hines and R. E. Pugh, "The spatial distribution of signal sources in meteoric forward scattering, Can. Jour. Phys., 34, 1005 (1956).

20. J. S. Greenhow and E. L. Neufeld, "The diffusion of ionized meteor trails in the upper atmosphere," J. Atmosph. Terr. Phys., 6, 133 (1955).

21. D. W. R. McKinley and A. G. McNamara, "Meteoric echoes observed simultaneously on back scatter and forward scatter," Can. Jour. Phys., 34, 625 (1956).

22. P. A. Forsyth and E. L. Vogan, "The duration of forward-scattered signals from meteor trails," Can. Jour. Phys., 34, 535 (1956).

23. J. M. Taff and J. Damelin, "A preliminary survey of meteoric reflection of vhf signals recorded at Allegan, Michigan, during August and September, 1951," F.C.C. report T.R.D. 2.1.5.

24. P. A. Forsyth and E. L. Vogan, "Forward-scattering of radio waves by meteor trails," Can. Jour. Phys., 33, 176 (1955).

25. H. J. Wirth, "Preliminary observations of forward scattering of electromagnetic waves by meteor trails," Naval Electronics Laboratory Research Report No. 690, 9 May, 1956.

Code	Organization	No. of Copies
AF 5	Commander Air Force Missile Test Center Patrick Air Force Base, Florida ATTN: Technical Library	1
AF 17	Commander Air Research and Development Command P. O. Box 1395 Baltimore 3, Maryland ATTN: RDTDR	1
AF 18	Director Air University Library Maxwell Air Force Base Alabama ATTN: CR-4582	1
AF 33	Commander Rome Air Development Center Griffiss Air Force Base New York ATTN: RCSST-3	2
AF 43	Flight Res. Laboratory, Res. Division Wright Air Development Center Wright-Patterson Air Force Base Dayton, Ohio ATTN: WCRRA	1
AF 68	Commander Wright Air Development Center Wright-Patterson Air Force Base Dayton, Ohio ATTN: Mr. Paul Springer WCLRE-5	1
AF 91	Office of Scientific Research Air Research and Development Command P. O. Box 1395 Baltimore 3, Maryland	2
AF 121	Commander Rome Air Development Center Griffiss Air Force Base New York ATTN: Mr. Charles A. Strom, RCEC	1
AF 128	Chief, Air Weather Service MATS, USAF ATTN: DSS Washington 25, D. C.	1
Ar 5	Signal Corps. Engineering Laboratories Tech. Documents Center Evans Signal Laboratory Belmar, New Jersey	1
Ar 7	Office of Chief Signal Officer Signal Plans and Operations Div. ATTN: SIGOL-2, Room 20 Com. Liaison Br., Radio Prop. Section The Pentagon, Washington 25, D. C.	1
Ar 9	Department of the Army Office of the Chief Signal Officer Washington 25, D. C. ATTN: SIGGD	1
Ar 28	Asst. Sec. of Defense for Research and Development Information Office Library Branch Pentagon Building Washington 25, D. C.	2
Ar 29	Technical Information Officer Signal Corps Engineering Laboratory Fort Monmouth, New Jersey	1
Ar 52	Director Evans Signal Laboratory Belmar, New Jersey ATTN: Mr. McAfee	1
G 2	Armed Services Tech. Info Agency Document Service Center Knott Building Dayton 2, Ohio ATTN: DSC/SA	10
G 11	Central Radio Prop. Laboratory National Bureau of Standards Boulder, Colorado ATTN: Technical Reports Library	1
G 18	Coordinating Committee on General Sciences Office of the Assistant Secretary of Defense Room 3 D-137, The Pentagon Washington 25, D. C.	1
G 22	Federal Communications Commission Washington 25, D. C. ATTN: Mr. Wm. C. Boese	1
G 25	Upper Atmosphere Research Section Central Radio Propagation Laboratory National Bureau of Standards Boulder, Colorado	1
G 26	Central Radio Propagation Laboratory National Bureau of Standards Boulder, Colorado ATTN: K. A. Norton	1
I 2	Stanford Research Institute Menlo Park, California ATTN: Dr. J. V. N. Granger, Head, Radio Systems Lab.	1
I 11	Bell Telephone Labs 463 West Street New York 14, N. Y. ATTN: Dr. S. O. Rice	1
I 53	Hughes Aircraft Company, Florence and Teal Streets Culver City, California ATTN: M. Bodner Research and Development Library	1
I 103	Stanford Research Institute Menlo Park, California ATTN: Dr. Peterson	1
I 116	Melpar, Inc. 3000 Arlington Boulevard Falls Church, Virginia ATTN: Mr. C. B. Raybuck Vice-President and Chief Engineer	1
I 128	Raytheon Manufacturing Company Services Building Waltham 54, Mass. ATTN: Mr. Whitcraft	1
I 132	Bell Telephone Laboratories 463 West Street New York 14, New York ATTN: Mr. K. Bullington	1

Code	Organization	No. of Copies
I 133	Collins Radio Company 855-35th Street, N. E. Cedar Rapids, Iowa ATTN: Irving Gerks	1
I 198	RCA Laboratories Princeton New Jersey ATTN: Mr. Richard Jenkins	1
I 220	Raytheon Manufacturing Company 100 River Street Waltham 54, Mass. ATTN: Dr. K. C. Black	1
I 221	Federal Telecommunication Labs. 500 Washington Avenue Nutley, New Jersey ATTN: Mr. Paul Adams	1
I 226	Rand Corporation 1700 Main Street Santa Monica, California ATTN: Dr. Edward E. Reinhart	1
I 229	Pickard and Burns 240 Highland Avenue Needham 94, Mass. ATTN: Dr. DeBettencourt	1
M 5	CROOTR-2	2
M 6	CROOTR-2 E	5
N 1	Chief, Bureau of Aeronautics Department of the Navy Washington 25, D. C. ATTN: Aer-TD-414	2
N 14	Office of Chief of Naval Operations Department of the Navy Op-533 Washington 25, D. C.	1
N 17	Department of the Navy ATTN: Mr. R. S. Baldwin, Code 835A Room 3315, Washington 25, D. C.	1
N 19	Director U. S. Naval Electronics Laboratory Point Loma, San Diego 52, California	1
N 27	Librarian U. S. Naval Postgraduate School Monterey, California	1
N 28	Air Force Development Field Representative Naval Research Laboratory Code 1110 Washington 25, D. C.	1
N 29	Director Naval Research Laboratory Washington 25, D. C. ATTN: Code 2021	2
N 34	Naval Research Laboratory Washington 25, D. C. ATTN: Dr. Leslie G. McCracken, Jr. Code 3935A	1
N 36	Office of Naval Research Department of the Navy ATTN: Geophysics Branch, Code 416 Washington 25, D. C.	1
N 45	Naval Electronic Laboratory San Diego, California ATTN: D. F. Heritage	1
N 51	Commanding Officer and Director U. S. Naval Underwater Sound Laboratory Fort Trumbull, New London, Connecticut	1
N 66	Chief of Naval Research Electronics Branch (Code 427) Department of the Navy Washington 25, D. C. ATTN: Dr. Arnold Shostak	7
N 67	Director Naval Research Laboratory Technical Information Officer (Code 2000) Washington 25, D. C.	1
N 68	Director Naval Research Laboratory ATTN: Code 5270 Washington 25, D. C.	1
N 69	Commanding Officer ONR Branch Office 346 Broadway New York 13, New York	1
N 70	Commanding Officer ONR Branch Office 86 E. Randolph Street Chicago 1, Illinois	1
N 71	Commanding Officer ONR Branch Office 1030 E. Green Street Pasadena 1, California	1
N 72	Commanding Officer ONR Branch Office 1000 Geary Street San Francisco 9, California	1
N 73	Commanding Officer Office of Naval Research Navy No. 100 Fleet Post Office New York, N. Y.	1
N 74	Chief, Bureau of Ships (810) Department of the Navy Washington 25, D. C.	1
N 75	Chief, Bureau of Ordnance (Re) Department of the Navy Washington 25, D. C.	1
N 76	Chief, Bureau of Aeronautics (El) Department of the Navy Washington 25, D. C.	1
N 77	Commander Naval Air Development and Material Center Johnsville, Pennsylvania	1

Code	Organization	No. of Copies
U 14	Research Prof. of Aerological Eng. College of Engineering Dept. of Electrical Engineering University of Florida Gainesville, Florida ATTN: Dr. Sullivan	1
U 16	Georgia Tech. Research Institute 225 N. Avenue, N. W. Atlanta, Georgia ATTN: Dr. James E. Boyd	1
U 23	Applied Physics Laboratory Johns Hopkins University 8621 Georgia Avenue Silver Springs, Maryland	2
U 27	Mass. Institute of Technology Lincoln Lab. P. O. Box 390 Cambridge 39, Mass. ATTN: Mr. Radford	1
U 34	Research Laboratory of Electronics Mass. Institute of Technology Room 20B-221 Cambridge 39, Mass.	1
U 37	University of Michigan Engineering Research Institute Willow Run Laboratories Willow Run Airport, Ypsilanti, Michigan ATTN: Librarian	1
U 49	Stanford University Electronic Research Laboratory Stanford, California ATTN: Dr. Eshleman	1
U 63	Cornell University School of Electrical Engineering Ithaca, New York ATTN: Dr. W. Gordon	1
U 64	University of Texas Electrical Eng. Research Lab. Box 8026, University Station Austin 12, Texas ATTN: Prof. A. W. Straiton	1
U 65	Massachusetts Institute of Technology Research Laboratory of Electronics Cambridge, Mass. ATTN: Prof. Wiesner	1
U 82	Massachusetts Institute of Technology Navy Representative, Lincoln Laboratory P. O. Box 73 Lexington 73, Mass.	1
U 83	Control Systems Laboratory University of Illinois Urbana, Illinois ATTN: Dr. F. W. Loomis, Director	1
U 84	Electronics Research Laboratory Stanford University ATTN: Dr. O. G. Villard Stanford, California	1
U 85	Department of Electrical Engineering University of Tennessee ATTN: Dr. F. V. Schultz Knoxville 16, Tenn.	1
U 86	Department of Electrical Engineering University of Florida ATTN: Prof. M. H. Latour Gainesville, Florida	1
U 94	Ohio State University Columbus, Ohio ATTN: Prof. C. T. Tai Dept. of Electrical Engineering	1
G 46	Ionospheric Research Section Central Radio Propagation Laboratory National Bureau of Standards Boulder, Colorado ATTN: R. C. Kirby, Chief	1
AF 157	Wright Air Development Center Wright-Patterson Air Force Base Dayton, Ohio ATTN: Mr. L. Hallman	1
Ar 32	Signal Corps Radio Propagation Agency Fort Monmouth, New Jersey ATTN: Mr. Fred Dickson, SCRPA	1
U 81	Harvard College Observatory 60 Garden Street Cambridge, Mass. ATTN: Mr. Gerald Hawkins	1

www.ingramcontent.com/pod-product-compliance
Lightning Source LLC
Chambersburg PA
CBHW082059210326
41521CB00032B/2531